*ZOSIMOS OF PANOPOLIS ON THE LETTER OMEGA*

Society of Biblical Literature

TEXTS AND TRANSLATIONS
GRAECO-ROMAN RELIGION SERIES

edited by
Hans Dieter Betz
Edward N. O'Neil

Texts and Translations Number 14
Graeco-Roman Religion Number 5
*ZOSIMOS OF PANOPOLIS ON THE LETTER OMEGA*
by
Howard M. Jackson

# ZOSIMOS OF PANOPOLIS
# ON THE LETTER OMEGA

edited and translated by
Howard M. Jackson

Scholars Press

# ZOSIMOS OF PANOPOLIS ON THE LETTER OMEGA
by
Howard M. Jackson

Copyright © 1978
Society of Biblical Literature

Library of Congress Cataloging in Publication Data

Zosimos of Panopolis.
    On the letter omega.

    (Graeco-Roman religion series ; 5) (Texts and
translations - Society of Biblical literature ; 14)
    Greek text and English translation of Peri tou
omega stoicheiou.
    Bibliography: p.
    Includes index.
    1. Alchemy. I. Title. II. Series. III. Series:
Society of Biblical literature. Texts and translations ;
14.
QD25.Z6713        540'.1        78-18264
ISBN 0-89130-250-6 pbk.

# TABLE OF CONTENTS

PREFACE

My work on Zosimos' *On the Letter Omega* began as a research paper presented to the members of the Corpus Hellenisticum seminars at the Claremont Graduate School in 1975. I owe a debt of gratitude to all the learned members of the group for their many perceptive suggestions, but the responsibility for any errors of fact or judgment be upon my own head! I wish to thank especially Dr. Edward N. O'Neil, professor of Classics at the University of Southern California, and Dr. Hans Dieter Betz, director of the seminar and gentlest of mentors, whose knowledge and patience saved me from many a blunder.

I have provided the reader with an introduction, Greek text and critical apparatus, English translation and notes, select bibliography and word-index to our tractate. I regret that I was not able to furnish it with a full commentary, because it cries for one, but the parameters of the Texts and Translations series forbid it. Still, the brevity of the tractate permitted the inclusion of what I hope are notes numerous and plenary enough to point up all the major problems of the text and the translation and to make Zosimos' meaning plain, or at least plainer, in every instance where the lack of a note might leave the reader puzzled.

Finally, though my work on Zosimos was a labor of love, I have learned through it to appreciate the wisdom of the Preacher, who found that "of making many books there is no end; and much study is a weariness of the flesh."

Howard Jackson
Ojai, California

April, 1978

INTRODUCTION

This little tractate by the Egyptian alchemist Zosimos
is of absorbing interest as a miniature showcase of the
religious and philosophical trends of the first centuries
of the Christian era. The obscurity in which it has lan-
guished is, however, hardly surprising. Largely scorned
by historians of science as rubbish, and gladly relegated
to their domain by humanists, the Graeco-Egyptian alchemical literature[1] has, on the whole, remained a virtual *terra
incognita* even to scholars of its cultural milieu. Yet
this body of literature is a valuable source of information
for the study of the many elements that, much like some
strange amalgam of the sacred art itself, fused to produce
the complex, eclectic thought of late antiquity.

In this respect Zosimos' *Authentic Commentaries on
the Letter Omega* is among the most instructive documents
of the alchemical corpus. Our author begins with a polemic
directed against a rival school of alchemists; the bone of
contention between them is the relevance of a procedure
for tincturing recommended by an old treatise on furnaces.
This treatise enjoined the observance of astrologically
propitious occasions in effecting tinctures, but Zosimos'
opponents, swelled-headed from their success, have ridiculed these precautions as unnecessary. When their formulas fail, however, they are driven to confess the validity
of the treatise's stipulations, only to revert to their
former ridicule again when their *daimon* grants them re-
newed success (sections 2 and 3).

Zosimos naturally views these persons as dominated by
Fate, mistress of the material universe, drawn as they are
in her train now hither, now yon. He proceeds to apply to
his opponents a Hermetic characterization of Fate-bound
men as "mindless": they have no conception of anything beyond the material. They are not philosophers, for the

1

latter, Hermes and Zoroaster agree, are superior to Fate
and are not swayed either by her gifts or by her misfor-
tunes (sections 4 and 5).  An allegorical interpretation
of Hesiod's myth of Prometheus, Epimetheus, and Pandora,[2]
the guileful gift of Zeus, serves to support his warning
against blind devotion to the deceitful pleasures of a
happy Fortune (section 6).

   Philosophy is the way to transcendence of Fate.
Zosimos adduces two opposing[3] methods of overcoming her
seductive wiles: Zoroaster boasts of mastering the force-
ful use of magic upon Necessity, but Hermes, in a passage
saturated with Stoic theologoumena,[4] enjoins self-
understanding and a contemplation of the material universe
penetrated by the divine Mind in which men share and by
which they may be enlightened and return to their original
incorporeal state (section 7).

   The motif of escape from corporeality subject to Fate
under the guidance of the Hermetic savior-revealer Mind
prompts Zosimos to turn to speculation on the first man.
The interpretation of Prometheus as the primal Man who has
fallen and who has become Epimetheus, the slave of fleshly
desires (section 6; cf. section 12), makes the transition
natural.  Among the Greeks the first earthly man is Epimetheus, but to the Egyptians he is Thoth, while the
Chaldeans, Parthians, Medes and Hebrews call him Adam,
his name in the language of the angels.  Zosimos sets out
for us a variety of etymologies for this heavenly name,
and offers us, to account for the source of this information, a strange Egyptian version of the legend of the
Septuagint: the high priest of Jerusalem sent Hermes to
translate the Hebrew scriptures into Greek--and Egyptian
(sections 8 and 9)!

   Zosimos goes on to adduce the opinions of the Gnostic
seer Nikotheos as to the name of the primal, spiritual Man
within the fleshly Adam, his temptation and fall into embodiment, and his consequent enslavement to Fate at the
hands of her archontic planetary agents (sections 10 and
11).  Nikotheos has already been mentioned in the preface

attached to our tractate (section 1).  There Zosimos
adopts the seer's explanation, based upon Homeric references to Ocean, of the mystical significance of the letter
omega.

The following sections recount the rescue of the
primal Man and those who are part of him by the savior
Jesus Christ, who descends into this world to effect the
separation from embodiment (section 13).  The consummation
of the process takes place in an eschatological struggle
that involves opposition from a jealous Mimic *Daimon* and
the advent of his forerunner, a mysterious Antichrist os-
tensibly identified with the Persian Gnostic master Mani[5]
(section 14).  Hesiod's myth is pressed into service once
again to typify the fall and eventual conversion of the
spiritual man (section 16).

Such, briefly, is the content of our fascinating
tractate.  We may now ask: What do we know about Zosimos[6]
himself?  Unfortunately very little; our information derives for the most part from his own writings and from
those of later alchemists who cite him.  He was certainly
an Egyptian; the phrase "among us" in 9.1-2 of our tractate
is sufficient indication of this fact.  The title attached
to two of his other works describe him as a Theban,[7] but
other titles and authors are more explicit in making him a
Panopolitan.[8]  Panopolis, the City of Pan, was the Greek
name[9] for the present-day city of Akhmim on the east bank
of the Nile, an intellectual center in Coptic times.  Its
ancient Egyptian name was Chemmis.[10]

The late tenth century lexicon known as the Suda, *sub
voce* Ζώσιμος, calls him an Alexandrian philosopher.
Scholars have long supposed from this notice that Panopolis was Zosimos' birthplace, but that later he came to
reside at Alexandria.  We know from Zosimos himself[11]
that at least once he traveled to Lower Egypt (hence he
must at this time have lived and practiced his art in the
Thebais), for he informs his readers that he had examined
a furnace "in the ancient sanctuary of Memphis."  Now
since in the ancient list of alchemists attached to many

4

manuscripts of our corpus[12] Alexandria is entered along
with the temple of Memphis as Lower Egyptian localities
where the sacred art was practiced, it would be surprising
if he had not at some time taken up residence at the me-
tropolis.  The interest he shows in the Sarapeion and the
library of the Ptolemies in section 8 of our tractate
lends some support to this supposition.

About Zosimos' professional peregrinations we learn a
great deal more from an interesting report of a voyage he
made to Rome.  It occurs in a rather mutilated fifteenth
century Syriac manuscript in the library of the University
of Cambridge.  Among the thirteen ancient Greek alchemical
treatises in Syriac translation which the manuscript contains, the seventh is a reference manual compiled by Zosimos.  At the end of the catalogue of mineralogical entries
Zosimos gives us a rather confusing first-person account
of his trip, but this much, at least, is clear: he traveled to sites in Coele-Syria, on Cyprus and Lemnos, places
renowned for wonders of mineralogical interest.  He toured
Macedonia and Thrace, traveled on to Rome--he never informs
us of the purpose of his visit--and eventually made his way
back to Egypt again.[13]  This journey does not confirm a
residency at Alexandria, but it marks a mobility, charac-
teristic of the age, that makes it quite probable.

None of our sources provides us with an absolute
dating for any point in Zosimos' life.  Consequently his
*floruit* must be deduced from the character of his writings
themselves, from the traditions which he cites, and from
the commentators who, in turn, cite him as an authority.
Berthelot[14] dates him towards the end of the third century,
and so too does Festugière, though later he widens this
dating to include the beginning of the fourth century.[15]
The latter scholar's dating depends upon Riess' arguments[16]
in its favor; they may be tabulated as follows: (1) the
alchemist Synesios[17] cites Zosimos, though without explicitly naming him; Zosimos must therefore well antedate A.D.
389, for Synesios addresses his tractate "to Dioskoros,

priest of the great god Sarapis in Alexandria," and the Alexandrian Sarapeion was destroyed in that year.[18]   (2) Zosimos himself cites Porphyry.[19]  He must therefore be later than the last decades of the third century.  Finally, (3) the Gnostic source from which Zosimos quotes in section 14 of our tractate itself dates circa A.D. 290, for the solution to the riddle there posed on the name of the Mimic *Daimon*'s forerunner is all but certainly Μανιχαῖος (see note 72 ad loc. in our translation).  Since Mani was put to death by Bahram I (A.D. 273/4-276/7), this solution to the riddle provides us with a certain *terminus a quo* for the document from which Zosimos derived this information, and thus for Zosimos himself.[20]

In addition, Riess sees in the statement that after seven periods the Mimic *Daimon* would himself[21] come (14.12-13) a chiliastic reference that presupposes that three periods--that is, three generations--have already passed since the advent of the forerunner (for Mani circa A.D. 242), and concludes from this that Zosimos must have lived at the beginning of the fourth century.  This dating is supported by the likelihood that Zosimos' *floruit* would have followed upon the disastrous political and economic conditions in Egypt at the end of the third century.  Diocletian's monetary reforms, the widespread discontent that spawned Domitius Domitianus' revolt, and the upheaval created by the emperor's investiture of Alexandria in A.D. 296/7, would be events of the recent past.[22]  Similarly, the hatred of the Manichaeans displayed by Zosimos' source document would be most understandable if it were seen as influenced by Diocletian's violent edict against the sect promulgated circa A.D. 303.

The Suda records as the work of Zosimos, besides a *Life of Plato* (compare 8.2 of our tractate), "*Chemical Matters* addressed to his sister Theosebeia"[23] and remarks: "it consists of 28 books, each superscribed with a letter of the alphabet; some entitle it *Manufactures*."  This work was evidently a compendious alchemical manual compiled by

Zosimos himself[24] in order to coordinate and systematize
what was certainly a body of writings far more massive
than what survives from his hand. Each section superscribed with a letter of the alphabet treated a specific
facet of the art.[25] Our tractate, for example, and perhaps others that fit the general category of apparatus and
furnaces, fell under the letter omega. Section 1 of our
tractate would perhaps, then, have been a preface added to
the body of the work at the time that Zosimos made the
compilation, for it is clear that the greeting with which
the second section opens marks the beginning of the original correspondence with Theosebeia. The preface afforded
Zosimos the opportunity for speculation on the secret
meaning of the letter omega that headed this section of
the compendium, and in turn provided the foundation for
the subsidiary title given our tractate by later Byzantine
collectors of the alchemical corpus.

How exactly the 24 (or 26, counting the digamma and
the qoppa used in enumeration) letters of the Greek alphabet were distributed among the Suda's 28 books we do not
know. Scholars have offered many schemes in solution to
the problem, but it is still a matter of conjecture.[26]
The most reasonable solution, however, is to hold that the
compendium included material independent of alphabetic
superscription. It would have been in the nature of re-
dactional composition that Zosimos added to the compendium
of his writings at the time of its organization. We have
argued that such was the case with the preface to our
tractate. To make up the difference between the Suda's
books and the letters of the Greek alphabet one need only
posit the Τελευταία ᾿Αποχή, a sort of epilogue to the
compendium. Its title shows that it existed originally in
more than one book, though only one extract from Book One
survives.[27]

The full text of our tractate was first printed in
the edition of the corpus published in 1887/8 in three
volumes by the French chemist Marcellin Berthelot, with

the collaboration of Charles-Émile Ruelle, under the title
*Collection des Anciens Alchimistes Grecs*. For this edition I have used Ruelle's text, but I have emended it
wherever I thought it necessary in the light of all the
critical appraisals of our document known to me.[28]  Only
the first ten of the 19 sections into which Ruelle divides
our text are included in the present work; the rest of
the tractate is either of a strictly technical nature or
of an interest peripheral to the present purpose.[29]  I
have also redivided the text into sections in an effort to
provide the reader with a better indication of the periods
into which the train of thought naturally falls.[30]

Ruelle transcribed his text from a single manuscript
in the library of Saint Mark in Venice (Ruelle's M), the
oldest known manuscript of tractates from the alchemical
corpus.[31]  He collated it with one manuscript selected
from the family of Parisini (Ruelle's K); he chose it because, where M differs from K, a contemporary hand has
recorded the variant reading from M in the margin.  Fresh
consultation of all the manuscripts would repay the effort,
and that for several reasons.  First, the text is often
corrupt and consequently quite insecure (especially in
section 13), and, second, this troublesome fact is com-
pounded by the carelessness with which Ruelle has edited
it for Berthelot's third volume.  Third, and more importantly still, Berthelot himself consulted only a small
portion of all the manuscripts of the corpus scattered
throughout the great libraries of Europe that are known to
contain our tractate.[32]  That is a task that must await
the future.

[1]For a concise treatment of the subject the reader
may refer to the article "Alchemie" by Ernst Riess for
Pauly-Wissowa's *Realencyclopädie der classischen Altertums-
wissenschaft* I (Stuttgart: J. B. Metzlerscher Verlag, 1893)
cols. 1338-55. See also Marcellin Berthelot, *Les Origines
de l'Alchimie* (Paris: G. Steinheil, 1885); André-Jean Festugière, *La Révélation d'Hermès Trismégiste* I (Paris: J.
Gabalda, 1950) pp. 217ff., and the same author's article
"Alchymica" in *L'Antiquité Classique* 8 (1939) pp. 71ff.
Another useful article is that by Robert Forbes, "Chemie,"
for the *Reallexikon für Antike und Christentum* II (Stuttgart: Hiersemann Verlag, 1954) cols. 1061-73. His chapter
"The Origin of Alchemy" in *Studies in Ancient Technology* I
(Leiden: E. J. Brill, [2]1964) pp. 125ff., is a good intro-
duction to the subject in its wider near-Eastern context.

[2]Whence did Zosimos derive this allegorization? We
cannot enter here into a full discussion of the source-
critical questions posed by our tractate, but Festugière
(*La Révélation* I, p. 271, n. 10) rightly points to section
15 as offering an indication that Zosimos' sources were
principally two: "the sacred books of Hermes" and "the
Hebrews." Festugière regards the allegorical exegesis of
Hesiod's myth as a third, but evidence from the tractate
supports the supposition that it was included in Zosimos'
Hermetic sources. The first occurrence of the myth in
section 6 follows closely upon the Hermetic sections 4 and
5; the link is visible in the introductory διὰ τοῦτο in
6.1 and in the recurrence in the section of key ideas--
"gifts" and "good fortune"--from the passage that precedes
it. Moralizing interpretation of ancient Greek myths was
an exercise of Stoic philosophy from the outset, and the
distinctly Stoic flavor of the Hermetic sections 4, 5, and
7.4f. (see further n. 4 below) makes it natural to view
the exegesis of Hesiod's myth as an original part of Zosimos' Hermetic sources. Lastly, the sentence (12.9-11)
that closes the second section to concern itself with the
myth summarizes the second half of the Hermetic section 7,
in particular lines 12-18. One need not, however, presume
that the allegory was the creation of the Hermetist.

[3]For a good discussion of this point, see Festugière
in his essay "La Doctrine des 'Viri Novi' sur l'Origine et
le Sort des Ames" in *Mémorial Lagrange* (Paris: J. Gabalda,
1940) pp. 97ff.

[4]Late antiquity had become obsessed, if that is not
too strong a word, with Fate and with the possibility of
overcoming her noxious influence. The all-pervading popu-
larity of astrology and its deterministic cosmology is the

principal factor in this development. The Stoa's cosmology was itself, at least at first, totally deterministic, and its masters argued the problem of fate and providence, determinism and free will, for centuries. Adoption of astrology and an astral mysticism into the Stoic bosom occurred largely through the efforts and influence of Poseidonios. Hence it is not surprising to find Zosimos' Hermetic sources heavily Stoic in their treatment of Fate and escape from her sphere of influence in an inner self-sufficiency turned contemplative. The same is true of similar passages in tractates of our Hermetic corpus.

[5]Sections 10-11 and 13-14, as well as portions of section 1, derive from Nikotheos (see n. 3 to the translation), whose work would fall under Zosimos' second source group, "the Hebrews." The speculation on Adam's name in sections 8 and 9 may derive from him as well or, alternately, from other Jewish or Jewish-Christian apocryphal traditions that Zosimos knew of. Zosimos certainly knew the legend of the fallen angels, for Synkellos in the context of a discussion of the legend cites Zosimos as follows (*Chronographia* 13D-14A; I, 23.21-24.13 Dindorf): "The sacred scriptures, or books, my lady, say that there is a family of *daimons* that feels desire for (or, stronger, has sexual intercourse with) women. Hermes mentions them in his natural discourses; nearly every exoteric and esoteric treatise makes mention of them. This is what the ancient and divine scriptures say, that certain angels lusted after women and came down and taught them all the works of nature. As a result, he (Zosimos? Zosimos' source?) says, they fell (or, possibly, gave offense, viz., to God) and remained outside of heaven, because they taught men all that is wicked and of no profit to the soul. These scriptures also say that from them the giants were born. Their initial transmission of the tradition about these arts came from Chemes. He called this book the *Book of Chemes*, whence the art is called 'chemistry.'" Interestingly enough, Zosimos attests the dissemination of the myth by his time to circles on the fringe of Judaism--or outside of it altogether; we cannot assume that he knew it from *1 Enoch* or *Jubilees*. His interest in the myth is accounted for by the derivation of the alchemical art from the teachings of these heavenly beings passed on by their gigantic son Chemes. This tradition is far older than Zosimos; alchemical works had naturally been written under the name of this Titanic eponymous founder of the art, and they were in existence before Zosimos' time. He refers to Chemes, or Chumes, often elsewhere: see Berthelot III, xviii, 1, texte grec 169.9; xx, 2, texte grec 172.17; xxiv, 4, texte grec 182.18, and 7, texte grec 183.22 where he is called "a prophet." The alchemist Olympiodoros (Berthelot II, iv, 27, texte grec 84.12) refers to him as well. This alchemical legend also explains Zosimos' interest in allegorical exegesis of the Prometheus legend, in Nikotheos and the Hermetica. The descent or fall of the primal Man and

his rescue from enslavement in the material cosmos is an essential part of most Gnostic systems and of the Hermetic cosmogony in *Poimandres*.

[6]The name occurs with some frequency in late antiquity. The earliest attestation is the Thasian epigrammatist of the first century B.C. The best known Zosimos is perhaps the pagan historian of the early sixth century, with whom our Zosimos is not identical. The feminine form Ζωσίμη also occurs; see, for example, *Papyri Graecae Magicae* (ed. Preisendanz) LXVIII, 4-5 and 11.

[7]Berthelot III, xli, texte grec 211.14, and li, texte grec 239.2. Θηβαῖος is perhaps not an error, as Festugière thinks (*La Révélation* I, p. 262, n. 1), but an equivalent (Lower Egyptian?) for Θηβαίτης, a dweller in the Thebais.

[8]The titles are Berthelot III, ix, texte grec 143.19 with the note in the apparatus, and xi, texte grec 145.17. See also Berthelot IV, vi, 4, texte grec 274.7, and VI, iii, 3, texte grec 401.13. Synkellos introduces the quotation cited above in n. 5 by calling him "the Panopolitan philosopher" (*Chron.* 13D; I, 23.21-22 Dindorf) and Photios describes him as Θηβαῖος Πανοπολίτης (*Bibliotheke* 170).

[9]Strabo 17.813; Diodorus of Sicily 1.18.2 and 1.31.17; Pliny, *Nat. Hist.* 5.49.61; and elsewhere.

[10]Herodotus 2.91 and Plutarch, *De Iside et Osiride* 356D spell it Χέμμις; Diodorus 1.18.2 has Χεμμώ. This Upper Egyptian city is not to be confused with the city of the same name near Buto in the Delta mentioned by Herodotus 2.156.

[11]Berthelot III, xlvii, 1, texte grec 224.4-6.

[12]Berthelot I, ix, 3, texte grec 26.5-6.

[13]Marcellin Berthelot, *La Chimie au Moyen Âge* II (Paris: Imprimerie nationale, 1893) pp. 300-302.

[14]*Collection des Anciens Alchimistes Grecs* I (Paris: G. Steinheil, 1887) p. 201.

[15]The earlier opinion is given in "Alchymica," p. 75 and in his essay in *Mém. Lagr.*, p. 125; the later in *La Révélation* I, p. 239.

[16]In his article "Alchemie," Pauly-Wissowa, *Realencyclopädie* I, col. 1348. Richard Reitzenstein (*Poimandres* [Leipzig: B. G. Teubner, 1904] p. 9) agrees with Riess, though he regards Riess' deductions with regard to Mani from section 14 of our tractate as precarious.

[17]Berthelot II, iii, 9, texte grec 63.5.

[18]One might add that 8.10 of our tractate presumes its existence.

[19]Berthelot III, xxxii, texte grec 205.13-14.

[20]Since Egypt, and in particular Alexandria, was evangelized by the Manichaeans even before Mani's death, the *terminus a quo* could, on this basis, be even earlier than might otherwise be expected. See the Middle Persian Manichaean text M2, recto I, lines 9-33 in Geo Widengren, *Mani and Manichaeism* (London: Weidenfeld and Nicolson, 1965) pp. 27ff. and L. Ort, *Mani* (Leiden: E. J. Brill, 1967) p. 65.

[21]*Pace* Festugière who, along with other scholars, thinks that αὐτός in 14.14 refers to the Son of God and not to the Mimic *Daimon*; see *La Révélation* I, p. 271, n. 9.

[22]The opening section of the Τελευταία Ἀποχή (Berthelot III, li, 1, texte grec 239.4-240.8 and Festugière, *La Révélation* I, pp. 275-76) suggests that Zosimos worked under conditions of the strictest control by the Roman government.

[23]Compare 2.1-2 of our tractate. She was probably not, Reitzenstein argues (*Poimandres*, p. 266, n. 2), his actual sister, but a fellow member of a community, whether Hermetic, as Reitzenstein suggests, or alchemical, or both, whose participants addressed one another as brother and sister.

[24]Two references in another tractate from Zosimos' hand indicate that the alphabetic arrangement was made by the author himself, and not a later compiler. The first book of the Τελευταία Ἀποχή refers back to the sections under the letters kappa (Berthelot III, li, 10, texte grec 246.12-13) and omega (Berthelot III, li, 11, texte grec 246.21-22). This last reference appears to be to our tractate, but τῶν φώτων τὰς ἀποδείξεις that it informs Theosebeia she will find there are lacking in our tractate as it survives. Another tractate entitled *On Apparatus and Furnaces* (Berthelot III, xlvii), which may originally have formed a unit with ours, does not treat "fires" either. The tractate Berthelot III, lii, however, does. If this is the passage to which Zosimos is referring Theosebeia, then either these several tractates were originally a unit later broken up, or, more likely, each alphabetic section could encompass more than one separate work.

[25]The second treatise in the Cambridge Syriac alchemical manuscript referred to above contains long extracts from what is surely Zosimos' compendium. The treatise cites books on silver, copper, tin, mercury, lead, iron, electrum, stones, and other matters, generally with their

particular alphabetic rubric.  See Berthelot, *La Chimie* II, pp. 222ff.  Book VIII of this same treatise includes a citation of the same passage that Synkellos quotes from Zosimos (see n. 5 above), though not without some variation (ibid., p. 238).

[26]Reitzenstein (*Poimandres*, p. 267) conjectures that Zosimos added four of the supernumerary letters of the Coptic alphabet to the 24 of the Greek, but for this there is lacking any evidence from the sources.  His assumption that the digamma and the qoppa were not included in the Greek series appears, in the former case at least, to be mistaken.  The second treatise of the Syriac manuscript mentions Zosimos' Book VI on the working of copper under the digamma.  Festugière ("Alchymica," p. 75, n. 5, and *La Révélation* I, p. 263, n. 2) proposes to consider that Zosimos correlated the letters of the Greek alphabet with one of the seven planets in a series four times repeated.  But his arrangement creates further difficulties, and it is more likely that only the seven vowels were matched with the planets.  We have some indication of this system when omega is associated with Saturn in 1.1-2 of our tractate.  Riess ("Alchemie," cols. 1345-46) suggests the solution that I have adopted.  He offers a list of other works by Zosimos that may have been without alphabetic superscription in the compendium.

[27]Berthelot III, li, texte grec 239.1-2.

[28]Besides those works listed in n. 30 below, there are the following: Joseph Bidez and Franz Cumont print our sections 5 and part of 7 with a critical apparatus and notes on pp. 243-45 of *Les Mages Hellénisés* II (Paris: Société d'éditions "Les Belles Lettres," 1938), as fragment #0-99 for our tractate's references to Zoroaster. Arthur Nock and André-Jean Festugière in their edition of the *Corpus Hermeticum* IV (Paris: Société d'éditions "Les Belles Lettres," 1954) pp. 117-21, print the text of our sections 4, 5, 7, and part of 8 with a French translation and notes as their fragments divers ##19-21.  Julius Ruska (*Tabula Smaragdina* [Heidelberg: C. Winter, 1926] pp. 24-30) and Hans-Martin Schenke (*Der Gott "Mensch" in der Gnosis* [Göttingen: Vandenhoeck und Ruprecht, 1962] pp. 52-54) offer useful German translations.  But the only English translations known to me are translations of other translations--that by R. Hull in Carl Jung's *Psychology and Alchemy* (Princeton, NJ: Princeton University Press, [2]1968 from *Psychologie und Alchemie*, Zürich: Rascher Verlag, [2]1952) pp. 360-68, and that by Philip Mairet in Jean Doresse's *The Secret Books of the Egyptian Gnostics* (London: Hollis and Carter, 1960 from *Les Livres Secrets des Gnostiques d'Égypte*, Paris: Librairie Plon, 1958) pp. 100-101.  Neither of these is complete in terms of the text that it translates, nor very well translated to begin with, though Jung has some interesting comments.  Gilles Quispel, in his article "Der gnostische Anthropos und die

jüdische Tradition" (reprinted from *Eranos-Jahrbuch* 12
[1954] pp. 195-234, in his *Gnostic Studies* I [Istanbul:
Nederlands Historisch-Archaeologisch Institut in het
Nabije Oosten, 1974] pp. 173-95) paraphrases our sections
9-11.  Wilhelm Bousset (*Hauptprobleme der Gnosis* [Göttin-
gen: Vandenhoeck und Ruprecht, 1907] in particular on pp.
20-21, 49-50, 190-94, and 334) refers to our tractate and
follows Reitzenstein in discussing our text in its wider
cultural context.

[29]Further polemic against his opponents and an
address to Theosebeia, on which see n. 9 to the translation.

[30]Alexander Ferguson and Walter Scott (*Hermetica* IV
[Oxford: the Clarendon Press, 1936] pp. 104-10) reorder
Ruelle's sections 1-13.  Scott offers a lengthy and often
penetrating commentary on the text on pp. 112-53, and
Ferguson adds his comments in the Addenda to the volume
(pp. 484-86).  Scott's butchery of the text is a travesty
of textual criticism, and his redivision into sections is
consequently not much of an improvement over Ruelle.
Festugière adopts Scott's division for his French translation in *La Révélation* I (pp. 263-73) of Ruelle's sections 1-12 and the first sentence of 13.  His notes are of
excellent quality.  Reitzenstein (*Poimandres*, pp. 102-106
and 267) prints only a Greek text with critical notes, and
somewhat erratically follows his own occasional paragraphic
division of the text.

[31]See the notice of the codices, p. 59.

[32]They are enumerated, with a description of their
contents, in the volumes of the *Catalogue des Manuscrits
Alchimiques Grecs* (edited by Joseph Bidez, Franz Cumont,
and many others [Bruxelles: M. Lamertin, 1924 and following years]).

TEXT AND TRANSLATION

ΤΟΥ ΑΥΤΟΥ ΖΩΣΙΜΟΥ

ΠΕΡΙ ΟΡΓΑΝΩΝ ΚΑΙ ΚΑΜΙΝΩΝ

ΓΝΗΣΙΑ ΥΠΟΜΝΗΜΑΤΑ ΠΕΡΙ ΤΟΥ Ω ΣΤΟΙΧΕΙΟΥ

    1. τὸ ω στοιχεῖον στρογγύλον τὸ διμερές, τὸ
ἀνῆκον τῇ ἑβδόμῃ Κρόνου ζώνῃ κατὰ τὴν ἔνσωμον φράσιν·
κατὰ γὰρ τὴν ἀσώματον ἄλλο τί ἐστιν ἀνερμήνευτον ὃ
μόνος Νικόθεος <ὁ> κεκρυμμένος οἶδεν. κατὰ δὲ τὴν
5   ἔνσωμον τὸ λεγόμενον Ὠκεανός, θεῶν, φησί, πάντων
γένεσις καὶ σπορά· καθάπερ φησίν, αἱ μοναρχικαὶ τῆς
ἐνσώμου φράσεως. τὸ δὲ λεγόμενον μέγα καὶ θαυμαστὸν
ω στοιχεῖον περιέχει τὸν περὶ ὀργάνων ὕδατος θείου
λόγον καὶ καμίνων πασῶν μηχανικῶν καὶ ἁπλῶν καὶ
10  ἁπλῶς πάντων.

---

Tit. 2-3: verba ita dividunt Scott, Fest.; ΠΕΡΙ ΟΡΓΑΝΩΝ
    ΚΑΙ ΚΑΜΙΝΩΝ ΓΝΗΣΙΑ ΥΠΟΜΝΗΜΑΤΑ · ΠΕΡΙ ΤΟΥ Ω ΣΤΟΙΧΕΙΟΥ
    Ruelle, Ruska.
1.1  in margine M atramento cano ὁ λγ (λόγος coni. Ruelle)
    μῦθος scriptum sign. Ruelle. τὸ ante στρογγύλον add.
    Reitz., Fest.; στρογγύλον <ὂν καὶ> (τὸ exciso)
    διμερές Scott.
1.3  ὃ Reitz. et omn. edd.: ὁ Ruelle.
1.4  fort. leg. κεκρυμμένως Ruelle; acc. Bidez-Cumont;
    <ὁ> κεκρυμμένος Reitz., Fest. δὲ codd.; δὴ Fest.
1.9  καὶ ἁπλῶν secl. Ruelle.
1.10 πάντων corr. Fest. Zosimi tractatu (Berthelot text.
    gr. 247.1-2) collato tit. ΕΡΜΗΝΕΙΑ ΠΕΡΙ ΠΑΝΤΩΝ ΑΠΛΩΣ
    ΚΑΙ ΠΕΡΙ ΤΩΝ ΦΩΤΩΝ inscripto: πασῶν Μ.

BY THE SAME ZOSIMOS
ON APPARATUS AND FURNACES
AUTHENTIC COMMENTARIES ON THE LETTER OMEGA

1. Round Omega is the bipartite letter,[1] the one that in terms of material language[2] belongs to the seventh planetary zone, that of Kronos.[3] For in terms of the immaterial it is something else altogether, something inexplicable, which only Nikotheos the hidden[4] knows. In material terms Omega is what he calls "Ocean,[5] it says, 'the birth and seed of all gods,'" as he says, "the governing principles of material language."[6] What he calls[7] the great and wonderful letter Omega heads the section on apparatus for the liquid of sulphur,[8] furnaces of all sorts, mechanical and simple alike, and all matters in general.

2. Ζώσιμος Θεοσεβείη εὖ εἴη ἀεί.  αἱ καιρικαὶ
καταβαφαὶ, ὦ γύναι, εἰς χλευασμὸν ἐποίησαν τὴν περὶ
καμίνων βίβλον.  πολλοὶ γὰρ εὐμένειαν ἐσχηκότες παρὰ
τοῦ ἰδίου δαιμονίου ἐπιτυγχάνειν τῶν καιρικῶν ἐχλεύα-
5  σαν καὶ τὴν περὶ καμίνων καὶ ὀργάνων βίβλον ὡς οὐκ
οὖσαν ἀληθῆ.  καὶ οὐδεὶς λόγος αὐτοὺς ἀποδεικτικὸς
ἔπεισεν ὅτι ἀλήθειά ἐστιν εἰ μὴ αὐτὸς ὁ ἴδιος αὐτῶν
δαίμων, κατὰ τοὺς χρόνους τῆς αὐτῶν εἱμαρμένης μετα-
βληθείς, παραλαβόντος αὐτοὺς κακοποιοῦ δέ, εἶπεν.
10  καὶ τῆς τέχνης καὶ τῆς εὐδαιμονίας αὐτῶν πάσης
κωλυθείσης, καὶ ἐφ᾽ ἑκάτερα τραπέντων τῶν αὐτῶν τύχη
ῥημάτων, μόλις ἐκ τῶν ἐναργῶν τῆς εἱμαρμένης αὐτῶν
ἀποδείξεων ὡμολόγησαν εἶναί τι καὶ μετ᾽ ἐκείνων ὃν
πρότερον ἐφρόνουν.

---

2.1  εὖ εἴη ἀεί correxi: ευηειαει M; εὐήει ἀεὶ K; fort.
leg. χαίρειν Ruelle dubitanter; εὖ διάγειν Fest.;
εὐή<μερ>εῖ<ν> ἀεί Scott.  καιρικαὶ corr.: κεριкαὶ MK.
2.9  παραλαβόντος αὐτοὺς κακοποιοῦ δέ, εἶπεν: αὐτοὺς corr.
Scott, Fest.: αὐτοῦ M; δέ M: δαίμονος Scott, Fest.;
εἶπεν correxi: εἰπεῖν M; ἔπεισεν corr. Fest.; παρέ-
λαβεν αὐτόν, κακοποιοῦσαν δὲ εἰπών Ruska emendatione
Pfeiffer accepta.
2.14  ἐφρόνουν M: κατεφρόνουν Scott, Fest.

2. Zosimos to Theosebeia, may it be well with you always! The matter of tinctures to be effected at propitious times[9] has brought ridicule upon the book *On Furnaces,*[10] lady. For many persons, once they have acquired the favor from their *daimon* to succeed with these tinctures, have ridiculed even the book *On Furnaces and Apparatus* as not being true. And no argument, however much proof it offers, convinces them that it is true; they do not admit it unless their *daimon*[11] itself tells them so-- when it is transformed in the course of the changing times[12] of their fate[13] and a maleficent one takes them over. And so, when their skill and all their success[14] are frustrated, and the same formulas, whatever they might be, turn out first one way and then the opposite, then reluctantly, with clear proofs from their fate, they confess that there is some truth to it, even for those formulas they previously thought highly of.

3. ἀλλ' οἱ τοιοῦτοι οὐκ ἀποδεκτέοι οὔτε παρὰ
θεῷ οὔτε φιλοσόφοις ἀνθρώποις· πάλιν γὰρ τῶν χρόνων
σχηματισθέντων κατὰ τοὺς λεπτοὺς χρόνους καλῶς καὶ
τοῦ δαιμονίου σωματικῶς αὐτοὺς εὐεργετοῦντος, πάλιν
5 μεταβάλλεται ἐφ' ἑτέραν ὁμολογίαν, τῶν προτέρων
ἐναργῶν πραγμάτων πάντων λελησμένοι, πάντοτε τῇ
εἱμαρμένῃ ἀκολουθοῦντες καὶ εἰς τὰς λεγομένας καὶ
εἰς τὰ ἐναντία, μηδὲν ἕτερον τῶν σωματικῶν φανταζό-
μενοι, ἀλλὰ τὴν εἱμαρμένην.


4. τοὺς τοιούτους δὲ ἀνθρώπους ὁ ῾Ερμῆς ἐν τῷ
Περὶ φύσεων ἐκάλει ἄνοας, τῆς εἱμαρμένης μόνους
ὄντας πομπάς, μηδὲν τῶν ἀσωμάτων φανταζομένους, μήτε
αὐτὴν τὴν εἱμαρμένην τὴν αὐτοὺς ἄγουσαν δικαίως,
5 ἀλλὰ [τοὺς] δυσφημοῦντας αὐτῆς τὰ σωματικὰ παιδευτή-
ρια, καὶ τῶν εὐδαιμόνων αὐτῆς ἐκτὸς <μηδὲν> ἄλλο
φανταζομένους.

---

3.3 λεπτοὺς Μ: fort. leg. ἐκλεκτοὺς Ruelle.
3.5 μεταβάλλεται Μ: fort. leg. μεταβάλλονται Ruelle; acc.
Scott.
3.7 τὰς λεγομένας (scil. ὁμολογίας ut vid.) Μ: fort. leg.
τὰ λεγόμενα Ruelle; an τὰ <ὑφ' ἡμῶν> λεγόμενα? Scott.
4.2 ἄνοας Μ: fort. leg. ἄνους Ruelle.  μόνους Μ: μόνον
Reitz.
4.3 πομπάς Μ: fort. leg. πομπέας Ruelle.  μήτε Μ: μηδὲ
Reitz.
4.4 τὴν αὐτοὺς corr. Reitz.: τοὺς αὐτοὺς Μ.  αὐτὴν...
δικαίως Μ: <περὶ> αὐτὴν...δικαίους Keil apud Reitz.;
αὐτὴν...δικαίως <ὑπολαμβάνοντας> Scott.
4.5 τοὺς secl. Reitz., Fest., recte.
4.6 an εὐδαιμονιῶν? Fest.  μηδὲν ins. Reitz., Scott,
Fest.; cf. 16.7.

3. But people like this are unacceptable both to God
and to men of philosophy.[15] No sooner do the times[16] take
on favorable forms with the passage of minutes and their
*daimon* treats them to some material advantage, than they
change[17] to the other opinion again, forgetting all
the former clear proofs. They are always following Fate,
now to this opinion and then to its opposite. They have
no conception of anything other than the material; all
they know is Fate.

4. In his book *On Natural Dispositions* Hermes calls
such people mindless, only marchers swept along in the
procession of Fate, with no conception of anything incorporeal, and with no understanding of Fate herself, who
conducts them justly. Instead they insult the instruction
she gives through corporeal experience, and imagine nothing beyond the good fortune she grants.[18]

5. ὁ δὲ ῾Ερμῆς καὶ ὁ Ζωροάστρης τὸ φιλοσόφων
γένος ἀνώτερον τῆς εἱμαρμένης εἶπον τῷ μήτε τῇ
εὐδαιμονίᾳ αὐτῆς χαίρειν, ἡδονῶν γὰρ κρατοῦσι, μήτε
τοῖς κακοῖς αὐτῆς βάλλεσθαι, πάντοτε ἐναυλίαν
5  ἄγοντες, μήτε τὰ καλὰ δῶρα παρ᾽ αὐτῆς καταδεχόμενοι,
ἐπείπερ εἰς πέρας κακῶν βλέπουσιν.


6. διὰ τοῦτο καὶ ὁ ῾Ησίοδος τὸν Προμηθέα
εἰσάγει τῷ ᾽Επιμηθεῖ παραγγέλλοντα· τίνα οἴονται
οἱ ἄνθρωποι πασῶν μείζονα εὐδαιμονίαν; γυναῖκα
εὔμορφον, φησί, σὺν πλούτῳ πολλῷ· καὶ φησὶ μήτε
5  δῶρον δέξασθαι παρὰ Ζηνὸς ᾽Ολυμπίου, ἀλλ᾽ ἀποπέμπειν ἐξοπίσω, διδάσκων τὸν ἴδιον ἀδελφὸν διὰ
φιλοσοφίας ἀποπέμπειν τὰ τοῦ Διός, τουτέστι τῆς
εἱμαρμένης, δῶρα.

---

5.1   Ζωροάστρης corr. Ruelle (cf. ad 7.1): Ζωροάστρις MK
      (an ferendum?).
5.4   βάλλεσθαι M: βλάπτεσθαι coni. Scott, Zuretti apud
      Bidez-Cumont. ἐναυλίαν M, ἐναύλια K: ἐν αὐλίᾳ
      Bidez-Cumont ad αὐλίας (emendatione Kroll accepta)
      in 7.4 conf., Scott, Ruska, Fest.; ἐν <ἐν>αυλίᾳ
      Keil; acc. Reitz.; ἐναυλίαν (scil. ζωήν) L.S.&J.;
      ἡσυχίαν coni. Ruelle.
5.4-6 πάντοτε...ἄγοντας et ἐπείπερ...βλέπουσιν membra
      inter se permutat Fest.
5.5   ἄγοντες M: ἄγοντας Reitz., Scott, Bidez-Cumont,
      Fest. καταδεχόμενοι M fort. ferendum: καταδέχεσθαι
      Reitz., Bidez-Cumont, Fest.; 'possis et hoc et
      ἄγοντες (5.5) servare, interpunctione graviore post
      βάλλεσθαι statuta et lacuna post βλέπουσιν (5.6)
      signata' ait Fest.
5.6   κακῶν M: κακὸν Reitz.; κακὰ (i.e., τὰ δῶρα) Bidez-
      Cumont, Fest., *Mém.Lagr*, p. 125.
6.2-4 τίνα οἴονται...πλούτῳ πολλῷ secl. Reitz.
6.4   μήτε M: μή<πο>τε Ruelle ad Hesiod. *Opera et Dies* 86
      conf.

5. But Hermes and Zoroaster say that philosophers[19] as a class are superior to Fate because they neither rejoice in her good fortune, for they are master over pleasures, nor are they thrown by the evils she sends, as they always lead an inner life,[20] nor do they accept the fair gifts she offers, since they look to an end of ills.[21]

6. This is also why Hesiod[22] introduces Prometheus offering this advice to Epimetheus: "What do men consider a good fortune greater than all others?" "A shapely wife," he replies, "along with great wealth." And he[23] says "not to accept a gift from Olympian Zeus, but to send it back again," teaching his brother through philosophy to reject the gifts of Zeus, that is, of Fate.

7. Ζωροάστρης δὲ εἰδήσει τῶν ἄνω πάντων καὶ
μαγείᾳ αὐχῶν τῆς ἐνσώμου φράσεως φάσκει ἀποστρέφεσθαι
πάντα τῆς εἱμαρμένης τὰ κακὰ καὶ μερικὰ καὶ καθολικά.
ὁ μέντοι Ἑρμῆς ἐν τῷ Περὶ ἐναυλίας διαβάλλει καὶ τὴν
5   μαγείαν λέγων ὅτι οὐ δεῖ τὸν πνευματικὸν ἄνθρωπον τὸν
ἐπιγνόντα ἑαυτὸν οὔτε διὰ μαγείας κατορθοῦν τι, ἐὰν
καὶ καλὸν νομίζηται, μηδὲ βιάζεσθαι τὴν ἀνάγκην, ἀλλ᾿
ἐὰν ὡς ἔχει φύσεως καὶ κρίσεως, πορεύεσθαι δὲ διὰ
μόνου τοῦ ζητεῖν ἑαυτόν, καὶ θεὸν ἐπιγνόντα κρατεῖν
10  τὴν ἀκατονόμαστον τριάδα καὶ ἑᾶν τὴν εἱμαρμένην ὁ
θέλει ποιεῖν τῷ ἑαυτῆς πηλῷ, τουτέστι τῷ σώματι.
καὶ οὕτως, φησί, νοήσας καὶ πολιτευσάμενος θεάσῃ τὸν
θεοῦ υἱὸν πάντα γινόμενον τῶν ὁσίων ψυχῶν ἕνεκεν ἵνα
αὐτὴν ἐκσπάσῃ ἐκ τοῦ χώρου τῆς εἱμαρμένης ἐπὶ τὸν
15  ἀσώματον. ὅρα αὐτὸν γινόμενον πάντα - θεόν, ἄγγελον,
ἄνθρωπον παθητόν· πάντα γὰρ δυνάμενος πάντα ὅσα
θέλει γίνεται καὶ πατρὶ ὑπακούει διὰ παντὸς σώματος
διήκων· φωτίζων τὸν ἑκάστης νοῦν, εἰς τὸν εὐδαίμονα
χῶρον ἀνώρμησεν ὅπουπερ ἦν καὶ πρὸ τοῦ σωματικὸν
20  γενέσθαι, αὐτῷ ἀκολουθοῦντα καὶ ὑπ᾿ αὐτοῦ ὀρεγόμενον
καὶ ὁδηγούμενον εἰς ἐκεῖνο τὸ φῶς.

---

7.3     τὰ om. Reitz. (an fortuito? Fest.), Bidez-Cumont.
7.4     ἐναυλίας Keil; acc. Reitz.: ἀναυλίας codd.; ἀϋλίας
        corr. W. Kroll apud Pauly-Wissowa, *Realencyclopä-
        die* VIII, col. 799; acc. Ruska, Scott, Bidez-
        Cumont, Fest.; ἀναυδίας coni. Ruelle ad librum
        Hermeticum (*Corp. Herm.* XIII, ut vid.) tit. ΠΕΡΙ
        ΣΙΓΗΣ inscriptum conf. καὶ secl. Scott.
7.6,9   ἐπιγνόντα corr.: ἐπιγνῶντα codd.
7.6     οὔτε secl. Scott. κατορθοῦν corr.: καθορθοῦν codd.
7.7     καλὸν: κακὸν mavult Scott. μηδὲ codd.: μήτε Reitz.;
        οὐδὲ Scott.
7.11    θέλει corr. Ruelle: θέλειν codd. τῷ ἑαυτῆς πηλῷ Κ
        collatione sua Reitz.: τῷ ἑᾶν τῇ σπηλῷ Μ (sic
        saltem Ruelle); fort. leg. τῷ ἑᾶν τῷ πηλῷ Berthe-
        lot. τουτέστι Scott, Reitz., Bidez-Cumont: του-
        τέστιν codd., Fest.
7.15-16 ὅρα...παθητόν secl. Reitz. ut glossema Christianum.
7.17    καὶ πατρὶ ὑπακούει secl. Fest., *La Révél.* I, p.
        267, eadem ratione, ὅρα...παθητόν tamen ret.
7.18    διήκων <καὶ> Reitz., Bousset, Jung. ἑκάστης codd.,
        Scott, Jung: ἑκάστου coni. Ruelle; acc. Reitz.,
        Fest.; an ἑκάστης ψυχῆς? Reitz., recte puto.

7. Now Zoroaster boastfully affirms that by the knowledge of all things supernal and by the Magian science of the efficacious use of corporeal speech one averts all the evils of Fate, both those of individual and those of universal application. Hermes, however, in his book *On the Inner Life*,[24] condemns even the Magian science, saying that the spiritual man, one who has come to know himself, need not rectify anything through the use of magic,[25] not even if it is considered a good thing, nor must he use force upon Necessity, but rather allow Necessity to work in accordance with her own nature and decree. He must proceed through that one search to understand himself, and, when he has come to know God, he must hold fast to the ineffable Triad[26] and leave Fate to work what she will upon the clay that belongs to her, that is, the body. And with this way of thinking and of regulating one's life, he says, you will see the Son of God[27] become everything for the sake of holy souls, to draw her[28] up out of the realm of Fate into the realm of the incorporeal.[29] See him become everything! - god, messenger,[30] passible man;[31] for, being capable of everything, he becomes everything he so wills[32] and obeys the Father by pervading every body. He enlightens the mind of each soul and spurs it[33] on up to the realm of bliss, where it was even before it[34] was born into corporeality, following after him, and filled with yearning[35] by him, and guided by him into that light.[36]

8. καὶ βλέψαι τὸν πίνακα ὃν καὶ Βίτος γράψας,
καὶ ὁ τρίσμεγας Πλάτων καὶ ὁ μυριόμεγας ʽΕρμῆς, ὅτι
Θώυθος ἑρμηνεύεται τῇ ἱερατικῇ πρώτῃ φωνῇ ὁ πρῶτος
ἄνθρωπος, ἑρμηνεὺς πάντων τῶν ὄντων καὶ ὀνοματοποιὸς
5  πάντων τῶν σωματικῶν. οἱ δὲ Χαλδαῖοι καὶ Πάρθοι καὶ
Μῆδοι καὶ ʽΕβραῖοι καλοῦσιν αὐτὸν ʽΑδάμ, ᾧ ἐστὶν
ἑρμηνεία γῆ παρθένος καὶ γῆ αἱματώδης καὶ γῆ πυρρά
καὶ γῆ σαρκίνη. ταῦτα δὲ ἐν ταῖς βιβλιοθήκαις τῶν
Πτολεμαίων ηὕρηνται· ὃν ἀπέθεντο εἰς ἕκαστον ἱερὸν,
10  μάλιστα τῷ Σαραπείῳ, ὅτε παρεκάλεσεν ʽΑσενᾶν τὸν
ἀρχ<ιερέα> ʽΙεροσολύμων πέμψαντα ʽΕρμῆν ὃς ἡρμήνευσε
πᾶσαν τὴν ʽΕβραΐδα ʽΕλληνιστὶ καὶ Αἰγυπτιστί.

---

7.19 τοῦ corr.: τοῦτο codd., Scott; τοῦ τὸ corr.
Ruelle; acc. Jung (recte? cf. 14.7); τὸ secl.
Reitz., Fest.
7.20-21 an [ὑπʼ] αὐτοῦ ὀρεγόμενον καὶ <ὑπʼ αὐτοῦ> ὁδηγούμενον? Ferguson
8.1 καὶ Βίτος (καὶ βιτος ΜΚ) Reitz. et omnes: Κέβητος
emend. Ruelle, qui et Κέβης τε ἔγραψε coni.
γράψας codd.: ἔγραψεν Keil apud Reitz.
8.3 πρώτῃ secl. Reitz., Ruska, Fest., *La Révél.* I,
p. 268, Schenke: an ἱερατικῇ <καὶ> πρώτῃ? Fest.,
*Corp. Herm.* IV, p. 119.
8.4 <ὁ> ἑρμηνεὺς Reitz.
8.7 πυρρά corr. Scott: πυρά codd.
8.9 ὃν codd.: ὧν Reitz., Schenke.
8.10 παρεκάλεσεν codd.: παρεκάλεσαν emend. Reitz.
8.10-11 τὸν ἀρχ<ιερέα> ʽΙεροσολύμων corr. Ruelle: τῶν
ἀρχιεροσολύμων codd.
8.11 πέμψαντα ʽΕρμῆν ὃς ἡρμήνευσε (εἱρμήνευσε Ruelle
ex K, ut vid.: ἑρμήνευσε Μ): fort. leg. πέμψαντα
ʽΕρμῆν ὁ ἑρμηνεύσας Ruelle; πέμψαι [τὰ] ἑρμηνέα
emend. Scott, qui et πέμψαι ἑρμηνέας οἳ ἡρμήνευσαν
ponit; πέμψαντα ἑρμηνέα Fest.

8. Furthermore, look also at the tablet that Bitos[37] wrote, and Plato the thrice-great and Hermes the infinitely-great,[38] that in the original[39] hieratic language the first man, the interpreter of all that exists and the giver of names to all corporeal beings, is designated Thouthos. The Chaldeans, the Parthians, the Medes and the Hebrews call him **Adam,** which means "virgin earth," and "blood-red earth," and "fiery-red earth," and "fleshly earth." These interpretations are to be found in the libraries of the Ptolemies; they laid it[40] away in each sanctuary, above all in the Sarapeion, at the time when he[41] appealed to Asenas the high priest of Jerusalem to send Hermes, who translated[42] the whole of the Hebrew into Greek and Egyptian.

9. οὕτως οὖν καλεῖται ὁ πρῶτος ἄνθρωπος ὁ παρ᾽
ἡμῖν Θωὺθ καὶ παρ᾽ ἐκείνοις ᾽Αδάμ, τῇ τῶν ἀγγέλων
φωνῇ αὐτὸν καλέσαντες. οὐ μὴν δὲ ἀλλὰ καὶ συμβολικῶς
διὰ τεσσάρων στοιχείων ἐκ πάσης τῆς σφαίρας αὐτὸν
5 εἰπόντες κατὰ τὸ σῶμα. τὸ γὰρ ἄλφα αὐτοῦ στοιχεῖον
ἀνατολὴν δηλοῖ, τὸν ἀέρα· τὸ δὲ δέλτα αὐτοῦ στοιχεῖον
δύσιν δηλοῖ, τὴν <γῆν> κάτω καταδύσασαν διὰ τὸ βάρος·
...· τὸ δὲ μ<ῦ> στοιχεῖον μεσημβρίαν δηλοῖ, τὸ μέσον
τούτων τῶν σωμάτων πεπαντικὸν πῦρ τὸ εἰς τὴν μέσην
10 τετάρτην ζώνην.


10. οὕτως οὖν ὁ σάρκινος ᾽Αδὰμ κατὰ τὴν φαινο-
μένην περίπλασιν Θωὺθ καλεῖται· ὁ δὲ ἔσω αὐτοῦ
ἄνθρωπος ὁ πνευματικὸς καὶ κύριον ἔχει ὄνομα καὶ
προσηγορικόν. τὸ μὲν οὖν κύριον ἀγνοῶν διὰ τὸ τέως·
5 μόνος γὰρ Νικόθεος ὁ ἀνεύρετος ταῦτα οἶδεν. τὸ δὲ
προσηγορικὸν αὐτοῦ ὄνομα Φῶς καλεῖται, ἀφ᾽ οὗ καὶ
φῶτας παρηκολούθησε λέγεσθαι τοὺς ἀνθρώπους.

---

9.2  τῇ <γὰρ> Reitz.
9.3  fort. leg. καλέσασι Ruelle.
9.6  δηλοῖ <καὶ> Reitz.
9.7  δηλοῖ τὴν <γῆν> corr. Berth. apud add. et corr. (vol.
     III, p. 475 ad text. gr. 231.6): δηλοῖ τὴν codd.;
     δηλοῖ <καὶ> γῆν Reitz.; δηλοῖ <γῆν> τὴν Scott.
9.8  post βάρος lacunam stat. Reitz., Scott, Schenke, quam
     τὸ δὲ δεύτερον ἄλφα στοιχεῖον ἄρκτον δηλοῖ suppl.
     Scott. δηλοῖ <καὶ> Reitz.
10.2 αὐτοῦ secl. Bousset.
10.3 καὶ κύριον ἔχει ὄνομα corr.: ονι και κυρομα εχειον
     habent codd.
10.4 ἀγνοῶν codd.: ἀγνοῶ Reitz., Scott; fort. leg.
     ἀγνοοῦμεν εἰς τὸ τέως Ruelle.
10.5 ταῦτα: τοῦτο Scott.
10.6 Φῶς (ita Ruelle, Ruska) vel Φὼς (ita Reitz., Scott,
     Fest.).

9. So, then, the first man among us is named Thouth, and among them Adam, a name from the language of the angels.[43] And not only that, but with respect to the body the name they refer to him by is symbolic, composed of four elements[44] from the whole sphere. For the letter A of his name signifies the *ascendant* east, and *air*; the letter D of his name signifies the *descendant* west, and earth, which sinks *down* because of its weight;[45] ...;[46] and the letter M of his name signifies the *meridian* south, and the ripening fire in the *midst* of these bodies, the fire belonging to the *middle*, fourth planetary zone.[47]


10. So, then, the Adam of flesh is called Thouth with respect to the visible outer mould,[48] but the Man within him, the Man of spirit, has a proper name as well as a common one.[49] Now the proper name no one knows for the present, for only Nikotheos, the one who cannot be found,[50] knows it. But his common name is Phos, and from this it followed that men came to be known as "photes."[51]

11. ὅτε ἦν Φῶς ἐν τῷ παραδείσῳ διαπνεόμενος, ὑπὸ
τῆς εἱμαρμένης ἔπεισαν αὐτὸν ὡς ἄκακον καὶ ἀνενέργητον ἐνδύσασθαι τὸν παρ' αὐτῶν 'Αδάμ, τὸν ἐκ τῆς
εἱμαρμένης, τὸν ἐκ τῶν τεσσάρων στοιχείων.  ὁ δὲ διὰ
5    τὸ ἄκακον οὐκ ἀπεστράφη· οἱ δὲ ἐκαυχῶντο ὡς δεδου-
λαγωγημένου αὐτοῦ.


12. τὸν ἔξω ἄνθρωπον δεσμὸν εἶπεν 'Ησίοδος ᾧ
ἔδησεν ὁ Ζεὺς τὸν Προμηθέα.  εἶτα μετὰ τὸν δεσμὸν
ἄλλον αὐτῷ δεσμὸν ἐπιπέμπει, τὴν Πανδώραν ἣν οἱ
'Εβραῖοι καλοῦσιν Εὔαν.  ὁ γὰρ Προμηθεὺς καὶ 'Επι-
5    μηθεὺς εἷς ἄνθρωπός ἐστι κατὰ τὸν ἀλληγορικὸν λόγον,
τουτέστι ψυχὴ καὶ σῶμα.  καὶ ποτὲ μὲν ψυχῆς ἔχει
εἰκόνα ὁ Προμηθεύς, ποτὲ δὲ νοός, ποτὲ δὲ σαρκὸς διὰ
τὴν παρακοὴν τοῦ 'Επιμηθέως ἣν παρήκουσεν τοῦ Προ-
μηθέως τοῦ ἰδίου.  φησὶ γὰρ ὁ Νοῦς ἡμῶν· ὁ δὲ υἱὸς
10   τοῦ θεοῦ πάντα δυνάμενος καὶ πάντα γινόμενος, ὅτε
θέλει, ὡς θέλει φαίνει ἑκάστῳ.

---

11.1   Φῶς (ita Ruelle, Ruska) vel Φὼς (ita Reitz., Scott);
       <ὁ> φὼς Scott.  διαπνεομένῳ Keil; acc. Reitz., Ruska,
       Jung, Schenke; διατετμημένος ἀπὸ (pro ὑπὸ) Scott;
       διακηλούμενος? Ferguson.
11.2   ἔπεισαν <οἱ ἄρχοντες> Reitz.
11.3   αὐτῶν corr. Reitz.: αὐτοῦ codd.; ret. Jung.
11.4   τὸν corr.: τῶν in codd. scriptum sign. Reitz.
11.5   οἱ corr.: εἰ Ruelle ex codd., ut vid.
12.1   τὸν <γὰρ> Reitz.  ᾧ corr. Reitz.: ὃν codd.
12.2   μετὰ <τοῦτον> τὸν Ruelle; acc. Scott.
12.3   Πανδώραν corr.: Πανδώρην Ruelle ex codd.; ret. Scott.
12.4   γὰρ codd.: fort. leg. δὲ Ruelle.
12.7   ὁ Προμηθεύς secl. Reitz.
12.9   τοῦ ἰδίου <ἀδελφοῦ> Ruelle; τοῦ ἰδίου <νοῦ> Reitz.,
       Fest.; τοῦ ἰδίου <νοός> Scott: nihil est addendum.
       ὁ Νοῦς Reitz.: ὁ νοῦς Ruelle.
12.10  ὅτε codd.: ὅτι (scil. ὅ τι: sic Bousset) Reitz.,
       Schenke; ὅσα Scott.
12.11  φαίνει codd.: φαίνεται Scott (recte? cf. ἐφάνη 13.2).

11. When Phos was in the Garden, spirited along on the wind,[52] at the instigation of Fate they[53] persuaded him, since he was innocent and unactivated,[54] to clothe himself with their Adam, who comes from Fate, who comes from the four elements. But Phos, for his innocence, did not refuse, and they began to exult to think that he had been made their slave.

12. Hesiod called the outer man "fetter," with which Zeus bound Prometheus. Then, after this fetter, Zeus set yet another fetter upon him - Pandora,[55] whom the Hebrews call Eve. For Prometheus and Epimetheus are, by the allegorical method, a single man, that is, soul and body; and so Prometheus sometimes takes the form of soul, sometimes that of mind, and sometimes that of flesh because of Epimetheus' disregard when he disregarded the advice of Prometheus, his own foresight.[56] For our Mind[57] says: "The Son of God, being capable of everything and becoming everything, when he wills, as he wills, appears to each."[58]

13. Ἀδὰμ προσῆν Ἰησοῦς Χριστὸς <ὃς> ἀνήνεγκεν
ὅπου καὶ τὸ πρότερον διῆγον φῶτες καλούμενοι. ἐφάνη
δὲ καὶ τοῖς πάνυ ἀδυνάτοις ἀνθρώποις, ἄνθρωπος
γεγονὼς παθητὸς καὶ ῥαπιζόμενος. καὶ λάθρα τοὺς
5 ἰδίους φῶτας συλήσας, ἅτε μηδὲν παθών, τὸν δὲ
θάνατον δείξας καταπατεῖσθαι καὶ ἐῶσθαι. καὶ ἕως
ἄρτι καὶ τοῦ τέλους τοῦ κόσμου ἔπεισι λάθρα καὶ
φανερὰ συνὼν τοῖς ἑαυτοῦ, συμβουλεύων αὐτοῖς λάθρα
καὶ διὰ τοῦ νοὸς αὐτῶν καταλλαγὴν ἔχειν τοῦ παρ'
10 αὐτῶν Ἀδάμ, κοπτομένου καὶ φονευομένου, παρ' αὐτῶν
τυφληγοροῦντος καὶ διαζηλουμένου τῷ πνευματικῷ καὶ
φωτεινῷ ἀνθρώπῳ, τὸν ἑαυτῶν Ἀδὰμ ἀποκτείνουσι.

---

13.1    <ὃς> ins. Ruelle; <ὃς αὐτὸν> Reitz. προσῆν...
        ἀνήνεγκεν: fort. leg. πρώην...ἀνήνεγκε Ruelle.
13.1-2  Ἀδὰμ...καλούμενοι secl. Reitz., Fest., Schenke
        ut glossema Christianum.
13.2    τὸ secl. Reitz.
13.2-6  ἐφάνη...ἐῶσθαι secl. Reitz., Fest., Schenke eadem
        ratione.
13.4,7,8 λάθρα Reitz.
13.5    συλήσας corr. Ruelle: συλλήσας codd.
13.6    δείξας: fort. leg. δόξας Ruelle.
13.7    ἔπεισι emend. Reitz., Fest., Schenke: τόποισι
        codd.; sec. Jung; τοῖς μὲν emend. Scott, qui et
        <τοῖς δὲ> post λάθρα ins.
13.8    συνὼν coni. Reitz., Fest., Bousset, Schenke:
        συλλῶν codd.; fort. leg. συλλαλῶν Ruelle.
13.8-9  λάθρα καὶ secl. Scott.
13.9    καταλλαγὴν: ἀπαλλαγὴν emend. Scott. παρ' secl.
        Reitz., Scott.
13.10,12 κοπτομένου καὶ φονευομένου παρ' αὐτῶν et τὸν
        ἑαυτῶν Ἀδὰμ ἀποκτείνουσι secl. Reitz., Fest.,
        Schenke ut glossemata Christiana.

13. Jesus Christ drew nigh[59] to Adam and bore him up
to the place[60] where those named "photes"[61] dwelt before.[62]
And he also appeared to very powerless men by becoming a
man who suffered and was subjected to blows.[63] And he
secretly carried off as his spoil the "photes," who belong
to him, because he suffered nothing but instead showed
death trampled under foot and thrust aside. And both now
and until the end of the world he comes,[64] both secretly
and openly, to his own and communes with them by counsel-
ing them secretly and through their minds to get rid[65]
of their Adam. By cutting off and slaying their[66] Adam
whose guidance is blind and who is jealous of the Man of
spirit and light they kill their own Adam.

14. ταῦτα δὲ γίνεται ἕως οὗ ἔλθῃ ὁ ἀντίμιμος
δαίμων, διαζηλούμενος αὐτοῖς καὶ θέλων ὡς τὸ πρώην
πλανῆσαι, λέγων ἑαυτὸν υἱὸν θεοῦ, ἄμορφος ὢν καὶ
ψυχῇ καὶ σώματι. οἱ δὲ φρονιμώτεροι γενόμενοι ἐκ
5  τῆς καταλήψεως τοῦ ὄντως υἱοῦ τοῦ θεοῦ διδόασιν αὐτῷ
τὸν ἴδιον Ἀδὰμ εἰς φόνον, τὰ ἑαυτῶν φωτεινὰ πνεύματα
σώζοντες <εἰς> ἴδιον χῶρον ὅπουπερ καὶ πρὸ κόσμου
ἦσαν. πρὶν ἢ δὲ ταῦτα τολμῆσαι τὸν ἀντίμιμον, τὸν
ζηλωτήν, πρῶτον ἀποστέλλει αὐτοῦ πρόδρομον ἀπὸ τῆς
10 Περσίδος μυθοπλάνους λόγους λαλοῦντα καὶ περὶ τὴν
εἱμαρμένην ἄγοντα τοὺς ἀνθρώπους. εἰσὶ δὲ τὰ στοιχεῖα τοῦ ὀνόματος αὐτοῦ ἐννέα, τῆς διφθόγγου σῳζο-
μένης, κατὰ τὸν τῆς εἱμαρμένης ὅρον. εἶτα μετὰ
περιόδους πλέον ἢ ἔλαττον ἑπτὰ καὶ αὐτὸς ἑαυτοῦ
15 φύσει ἐλεύσεται.

---

14.2    διαζηλούμενος corr. Reitz.: δι᾽ οὗ ζηλούμενος
        codd. ὡς τὸ πρώην secl. Reitz. πρώην: πρῶτον
        Scott.
14.3    ὢν secl. Ruska, qui et ἄμορφος in ἄμομφος
        emendare vult.
14.4    φρονιμώτεροι γενόμενοι corr. Ruelle: φρονιμώτερον
        γενάμενοι codd.
14.5    διδόασιν corr. Reitz.: δίδωσιν codd.; δώσουσιν
        Scott.
14.7    <εἰς> ἴδιον χῶρον Reitz., Fest. ὅπουπερ: ὅπου
        Reitz., Fest.
14.11-12 in margine M prima manu scriptum: ση᾽ (sc.
        σήμαινε), cf. Ant.Lib. I.10-11, ed. I. Cazzaniga.
14.13   ὅρον: fort. leg. λόγον Reitz.
14.14   περιόδους corr.: περιόδου (Ruelle) vel περίοδον
        (Reitz.) codd. ἑαυτοῦ corr.: ἑαυτῷ codd.; <τῇ>
        ἑαυτοῦ Reitz.

14. And this will go on until the Mimic *Daimon*[67] comes, since he is jealous of them and wants to deceive them as before,[68] claiming that he is Son of God, ugly though he is both in soul and body. But because they have become wiser by their apprehension of Him who is truly Son of God, they yield up to him their Adam, yield him up to death, and save their shining[69] spirits by introducing them into their own realm, where they were even before the world.[70] But before the Mimic, the Jealous One, dares these things[71] he will first send his own forerunner from Persia, telling deceptive, fabulous tales and leading men on about Fate. The letters of his name are nine,[72] if the diphthong is preserved,[73] in accord with the pattern of Fate.[74] Then, after seven periods, more or less, he[75] will come in his own person.

15. καὶ ταῦτα μόνοι Ἑβραῖοι καὶ αἱ ἱεραὶ
Ἑρμοῦ βίβλοι περὶ τοῦ φωτεινοῦ ἀνθρώπου καὶ τοῦ
ὁδηγοῦ αὐτοῦ υἱοῦ θεοῦ καὶ τοῦ γηΐνου Ἀδὰμ καὶ τοῦ
ὁδηγοῦ αὐτοῦ ἀντιμίμου τοῦ δυσφημίᾳ λέγοντος ἑαυτὸν
5    εἶναι υἱὸν θεοῦ πλάνη.


16. οἱ δὲ Ἕλληνες καλοῦσι <τὸ>ν γῆΐ<ν>ον Ἀδὰμ
Ἐπιμηθέα συμβουλευόμενον ὑπὸ τοῦ ἰδίου νοῦ, τουτέστι
τοῦ ἀδελφοῦ αὐτοῦ, μὴ λαβεῖν τὰ δῶρα τοῦ Διός. ὅμως
καὶ σφαλεὶς καὶ μετανοήσας καὶ τὸν εὐδαίμονα χῶρον
5    ζητήσας, πάντα ἑρμηνεύει καὶ πάντα συμβουλεύει τοῖς
ἔχουσιν ἀκοὰς νοεράς· οἱ δὲ τὰς σωματικὰς ἔχοντες
μόνον ἀκοὰς τῆς εἱμαρμένης εἰσί, μηδὲν ἄλλο καταδεχόμενοι ἢ ὁμολογοῦντες.

---

15.1  <οἱ> Ἑβραῖοι?
15.4  λέγοντος corr.: λέγωντος codd.
15.5  <καὶ> πλάνη Keil; acc. Reitz.
16.1  καλοῦσι <τὸ>ν corr. Reitz.: καλοῦσιν codd. γῆΐ<ν>ον
corr.: γηΐον Ruelle ex codd.
16.5  post ζητήσας lacunam sign. Reitz., qui et ...<ὁ δὲ
Προμηθεὺς, τουτέστιν ὁ νοῦς suppl.; ita vel ...<ὁ δὲ
Νοῦς ἡμῶν> suppl. Fest.; nihil est addendum.
16.7  εἰσὶ <πομπαί> Reitz.; εἰσὶ <δοῦλοι> Scott.

15. And only the Hebrews and the sacred books of Hermes[76] say these things, about the Man of light and his guide, the Son of God, and about the Adam made of earth and his guide, the Mimic, who blasphemously claims to be Son of God in order to deceive.


16. But the Greeks call the Adam made of earth Epimetheus, who was counseled by his own mind, that is, by his brother, not to accept the gifts of Zeus. Since he[77] fell, however, and had a change of mind,[78] and searched for the realm of bliss, he explains all and counsels all for those who have ears of the mind. But those who have only bodily ears belong to Fate, for they neither grasp nor confess anything else.

[1]The shape of the Greek letter omega (ω) suggests the descriptions "round" and "bipartite." The reason for their inclusion is not so obvious, but it is probably in anticipation of correlating the letter omega with Ocean. Ocean was conceived to be a river that encircled the world; Homer describes the shield made by Hephaistos for Achilles as depicting Ocean flowing around its outer rim (*Iliad* 18. 607; cf. also Herodotus 4.8). This fact accounts for omega's being called στρογγύλον. The explanation for διμερές is perhaps that Zosimos held Ocean to be a hermaphrodite being. The alchemist Olympiodoros (who just may be identical with the Neo-Platonic commentator of Plato) cites Zosimos as saying that the sea is ἀρρενόθηλυς (Berthelot II, iv, 32, texte grec 89.19). The background for this odd, un-Greek conception is supplied by a statement in Diodorus of Sicily (actually his source Hekataios of Abdera): the Egyptians say that "the ancients named the moist element 'Okeane,' which means 'Sustenance Mother'; but by some of the Greeks it is held to be 'Okeanos' (i.e., masculine)" (1.12.5). In classical Egyptian cosmogonies the primeval waters of Chaos are a divine syzygy, Nwn and Nwnt. Furthermore, Diodorus (1.12.6) goes on to say that the Egyptians consider Ocean to be the Nile, and the ancient Egyptians often depicted the god Nile as a man with pendulous breasts.

[2]Literally, "according to embodied speech," that is, the language of corporeal humans as opposed to that of bodiless celestials. Compare the statement in 9.2-3 that "Adam" is a name derived from the language of the angels, and in 10.2-4 that the spiritual Man within the fleshly Adam has a proper name, which only Nikotheos knew, as well as a common one. The similarity of all these passages leads one to conclude that Zosimos derived them all from Nikotheos. See also 7.1-2.

[3]That is, Saturn, each of the seven Greek vowels being paired, in ascending order, with one of the seven planets. Kronos and Ocean were brothers, both sons of Heaven and Earth (Hesiod, *Theogony* 132-138), and after his rebellious son released him from bondage, Kronos ruled over the heroes who dwell on the Isles of the Blessed at the ends of the earth "along the shores of deep-swirling Ocean" (Hesiod, *Works and Days* 167-173; cf. *Theogony* 215-216). Clement of Alexandria (*Stromateis* 5.8.50.1) informs us that the Pythagoreans call the sea "the tears of Kronos."

[4]See also 10.5. Nikotheos was a (second century?) Jewish-Christian or perhaps secondarily Christianized Jewish visionary and prophet of some repute in Gnostic circles for his revelations of heavenly secrets acquired during journeys into that realm. He was evidently considered to have traveled there at or in lieu of death, and, as the epithets κεκρυμμένος here and ἀνεύρετος in 10.5, as well as διὰ τὸ τέως in 10.4, seem to intimate, Nikotheos was expected to inspire further revelations from his new home, and perhaps even eventually to return from his hiding place, probably in some eschatological context. The same traditions and expectations surround the figures of Enoch, Moses, and Elijah in intertestamental Jewish works of apocalyptic character, not to mention the New Testament. The work by or pseudepigraphically attributed to Nikotheos that Zosimos knew was plausibly the *Revelation of Nikotheos* mentioned by Porphyry along with those of Zoroaster, Zostrianos, Allogenes, Messos, and nameless others, in his *Life of Plotinus* 16. This lost apocalypse may originally have been Jewish (Sethian?) and only secondarily Christianized by Christian Gnostics (cf. sections 13 and 14 of our tractate), for Porphyry merely states that the Gnostic Christians against whom Plotinus taught boasted possession of it. The Nag Hammadi tractates *Zostrianos*, which also claims the authority of Zoroaster, and *Allogenes*, which often mentions Messos, are likely to be those very works named by Porphyry. They too are revelations of heavenly secrets, and they are not Christian. For translations of them see James Robinson, ed., *The Nag Hammadi Library* (New York: Harper & Row, 1977) pp. 368ff. and 443ff. Zostrianos and Allogenes are, like Nikotheos, seers to whom the revelations are made and who in turn reveal their secrets to the elect. Nikotheos is never mentioned in the Nag Hammadi codices; we know of his activities from the *Untitled Apocalypse* of the Coptic Gnostic codex Brucianus. There he is mentioned along with one Marsanes as one of the "perfect," one who had managed to speak--"with a tongue of flesh," it seems, though this is declared impossible--about the Father and his Monogenes who is Light (see 10.5-7 of our tractate). Nikotheos spoke about him; "he had seen him, for he is one who was there....He revealed the invisible perfect Triple-Power." See Charlotte Baynes, *A Coptic Gnostic Treatise contained in the Codex Brucianus* (Cambridge: the University Press, 1933) pp. 83-84 with note 7, which refers to Zosimos. Carl Schmidt, *Koptisch-Gnostische Schriften*, erster Band (Berlin: Akademie Verlag, [3]1962) pp. 341-342, translates "Nikotheos hat...ihn geschaut, denn er ist jener," i.e., still there. There is an exceedingly fragmentary tractate under the name *Marsanes* in the Nag Hammadi library; it too, like the others mentioned above, is a revelation of heavenly secrets and, to judge from the remains, not Christian either. See Robinson, *Nag Hammadi Library*, pp. 417ff.

[5]The connection is not only that in Greek the word "ocean" begins with an omega, but also that in late antiquity the word στοιχεῖον had come to designate any primal or fundamental world-originating principle, as well as a letter of the alphabet (so again in 14.11-12). Here the word is used of Ocean from the classical Greek cosmogonies; 1.1-2 suggests the planetary rulers and demiurges; and 9 and 11.4 the four physical elements earth, water, air and fire. For this developed use of the word στοιχεῖον see Gal 4:3 and 9 and Col 2:8 and 20, to cite but the New Testament.

[6]The meaning of καθάπερ...φράσεως is quite obscure; even Festugière (La Révélation I, p. 264, note 6) despairs. This translation understands 'Ωκεανός, θεῶν...σπορά and αἱ μοναρχικαὶ...φράσεως to be a citation by Zosimos from the Nikotheos source, to whom καθάπερ φησίν is Zosimos' reference, and that φησί is the cited document's reference to an interpolated Homer (cf. Iliad 14.201, 246, and 302). Homer was well enough known not to need to be explicitly named; one might compare the 'ipse dixit' of the Pythagoreans. The description αἱ μοναρχικαί would be Nikotheos' explanation of the function of γένεσις καὶ σπορά, both feminine.

[7]Or, less probably, simply "what is called," "the proverbial." So also in 1.7.

[8]Or, "the divine water." There is a play here, common elsewhere in the writings of the alchemists, upon the words θεῖος "divine" (as Ocean was in ancient myth) and τὸ θεῖον "sulphur," for which the alchemists used omega as a sign. See the list of alchemical symbols and shorthand notations prefixed to the manuscript Venetus Marcianus 299, our M, printed in Berthelot, Collection I, p. 104, fig. 3, right hand column, the third entry from the bottom. What exactly "liquid of sulphur" was chemically is obscure to me; it seems to be sulphur in some fusible state, and its production required some apparatus because, whatever it was, it was "smelted from lead" (cf. the alchemical lexicon Berthelot I, ii, sub voce θεῖον ὕδωρ, texte grec 8.13). Whatever the connection between sulphur and lead was, it was strong enough to warrant assignation of the same sign to both substances--the sign for the planet Kronos. See the alchemical sign-list of the manuscript Parisinus graecus 2327, Ruelle's A, reproduced by Berthelot, Collection I, p. 114, fig. 7, lines 11-12. Lead was Saturn's metal; see Berthelot, ibid., pp. 73ff. and, e.g., Origen's Contra Celsum 6.22. This "liquid of sulphur" or "divine water" effected the sublimation of metals from their ores; "it produces the transformation; by its application you will bring out what is hidden inside; it is called 'the dissolution of bodies'" (Berthelot II, iii, 6, texte grec 61.1-3). Hence the play on "divine." See also Berthelot II, iv, 38, 42 with notes

3 and 4, 43-44, 46-47; III, vii; viii; x, 2-3; xiv, 1;
xv, 1-4; xvi, 10, 13; xvii; xxi, 1-2; xxv; xxviii, 8;
lii, 2-4.

[9]That is, under the influence of favorable astrological conditions. See Ruska, *Tabula Smaragdina*, pp. 22-23
and Festugière, *La Révélation* I, p. 264, note 10, as well
as what follows here.

[10]The reference is not, as Festugière (*La Révélation*
I, p. 265, note 1) points out, to the present (Zosimos'
own) treatise on apparatus and furnaces, but to a treatise
already in existence. It may be identical with the καμι-
νογραφία of Maria the Jewess referred to by the alchemist
Olympiodoros (Berthelot II, iv, 35, texte grec 90.19).
The treatise was evidently of some age, for Zosimos refers
to it as a writing of the ancients, and demurs to Theosebeia's requests in her letters for information on apparatus by conceding his inability to treat the subject more
authoritatively than this older treatise has, though he
proceeds to set out briefly what he knows "in order that
we may understand all that they (viz., the ancients) said"
(Berthelot III, xlix, 13, texte grec 234.1-10, not included
in this edition). This explains the respect that Zosimos
shows for the treatise in the present section. Zosimos
refers to this treatise with a fuller title in 2.5.

[11]The conception of a personal genius in force here
is astrologically conditioned in that the *daimon* governing
the person and his success in enterprise--here alchemical--
is considered to be influenced in its nature by the particular configuration, now beneficent, now maleficent, of
astral forces in effect at any moment. See the following
note and 3.2-4 below.

[12]The word χρόνος as used here and in 3.2 below is a
technical astrological term and refers to the time taken
by each of the 360 degrees of the zodiacal constellations
to rise through ascendancy. The impact of each "time" was
colored by the particular planet or decan, called in this
respect the χρονοκράτωρ, whose power was aspectually domi-
nant for the period during which that degree was in the
ascendant. Zosimos' point is that his opponents only accept the validity of the procedures for καιρικαὶ καταβαφαί
set out in the book in question when their *daimon*, conditioned by or identical with a maleficent χρονοκράτωρ, im-
pedes the success they had formerly enjoyed with their own
methods under a beneficent one. When this latter condi-
tion returns, however, they revert to their former independent stance.

[13]Or, less personally, "of the Fate that is over
them"--i.e., that controls them--or even, sarcastically,
"of that Fate of theirs." In both these cases there would
be a glance forward to what Zosimos says of these people
in sections 3 and 4.

[14]The word play inherent in the use of εὐδαιμονία here, viz., with the proper meaning "the state of being blessed with a good *daimon*" and hence "happiness," "good fortune," "prosperity," cannot be preserved in English. Note the continuance of this theme in 4.6, 6.3, 7.18, and 16.4 below. The question of true happiness, one Fate-controlled versus Self-controlled, material versus spiritual, is the theme that links together all the various traditions that Zosimos cites in the rest of the tractate.

[15]I.e., alchemists, in this context. See also 5.1 and note 19.

[16]See note 12 above.

[17]Or perhaps "it (viz., these persons' *daimon*) changes their mind and they confess etc."

[18]I assume that the whole of this section is a citation from the Hermetic tract named in 4.2. The distinctly Stoic flavor of the passage argues in its favor; see note 2 to the Introduction, and with 4.6-7 compare 16.7-8, one of the Promethean sections. The passage 16.3-8 is Hermetic; it contains many close parallels to section 7. For parallels in the Hermetic corpus itself, see especially 4.5 and 7; 16.16; and *Asclepius* 7.

[19]See note 15. In late antiquity "philosopher" had come to encompass devotées of any mystical system of knowledge--Egyptian alchemists, Chaldean astrologers, Zoroastrian magicians; Neo-Platonists and theurgists; Gnostics and Hermetists: all were "philosophers." In fact, philosophy in the forms and traditions of classical Greece and religion in the late Roman period had totally coalesced.

[20]The meaning of ἐναύλιος is literally "within the courtyard gate" or "in the court," the force of the expression being that philosophers, by their realization of kinship with God through mind, the divine element awakened within them, pass the whole of their lives in the court of their Father, the divine King who is Mind. They are at home with themselves, there.

[21]The gifts only seem fair, as Zosimos--or rather his Hermetic source--goes on to illustrate from the myth of Pandora.

[22]See *Works and Days* 54ff., and the parallel passage *Theogony* 558ff. Only μήτε...ἐξοπίσω 6.4-6 approaches citation of Hesiod (*Works and Days* 86-88); the rest derives ultimately from moralizing Stoic propaedeutic allegorization. The identification of Zeus with Fate, for instance, was made already by Chrysippus in his book *On the Nature of the Gods* (Cicero, *De Natura Deorum* 1.14.40-41; see also 2.24.63f.). Vergil (*Aeneid* 3.375-376) and the

author of the Hermetic *Asclepius* (27) follow this tradi-
tional exegesis. Cicero tells us that Zeno himself taught
with respect to Hesiod's *Theogony* that Zeus and Hera and
all the names of the other gods refer *"per quandam sig-
nificationem* to dumb and lifeless things" (*De Natura
Deorum* 1.14.36).

[23]Prometheus.

[24]The text is insecure here. Most scholars accept
W. Kroll's emendation ἀϋλίας ("immateriality") for the
manuscripts' ἀναυλίας. In view of the mention of an
ἐναυλία (ζωή) in 5.4, however, it seems to me more plau-
sible, and simpler too, to correct to ἐναυλίας here, and
to consider sections 5 and 7.4ff. as citations from one
Hermetic tract. If a Hermetic treatise *On Immateriality*
were known, Kroll's suggestion would have greater weight.

[25]The term μαγεία, like the μάγος who practices it,
can mean either vile superstition and ground for oppro-
brium or theurgic power and mystic science, depending upon
the user's viewpoint. A passage in Apuleius' famous *Apol-
ogy* in defense of an accusation that he is a *magus* (25.5ff.)
is particularly enlightening in this regard. Zosimos pre-
sents us with two assessments of magic in regard to its
value in overcoming Fate. "Zoroaster," the Magus *par excellence*, not surprisingly endorses its use (7.1-3);
Apuleius (*Apol.* 25.11) cites pseudo-Plato's *First Alcibiades* 121E and maintains in Zoroaster's name that μαγεία is
"high-priestess of heavenly powers" (26.2). Zosimos' Hermetist condemns its use as needless (7.4-6) with a hint,
perhaps, of the unstated purpose of making a virtue of
legal necessity. In the eyes of Roman law, the practice
of magic was either itself a crime or a seedbed for a
variety of criminal activities, not the least of which was
*lèse majesté*. Zosimos' own evaluation seems to match
Zoroaster's; at least, the adverbial use of καί in 7.4
points in this direction. Egyptian respect for magic was
ancient and quite ineradicable. Other documents from the
Egypt of this period agree with Zosimos: the Hermetic extract 23 from Stobaeus (the *Kore Kosmou* 68) and a Berlin
magical papyrus (*Papyri Graecae Magicae*, ed. Preisendanz,
1.127).

[26]The Triad is evidently the Father, who is Mind, his
Son the *Logos* born of Mind (see the following note), and
either τὸ πᾶν, the material cosmos, or man as possessing
mind. Such triadic formulations are common in the Hermetica; for other examples, compare *Corpus Hermeticum*
8.2-3 and 5: God the Father, the All, and man; 13.18:
God, *Logos*, and the All; and *Asclepius* 10: God, world,
and man. In another tractate (Berthelot III, vi, 18,
texte grec 132.21-24) Zosimos illustrates an alchemical
procedure in which "a compositional monad results in an
inseparable triad" by remarking "and furthermore, a triad

in union which produces a triad in distinction constitutes
the world by the providence of the primal creative Cause
and Demiurge of creation.  That is why he (Hermes) is
called Thrice-greatest, because he envisioned the created
and the Creator triadically."  Trinitarian speculation was
not exclusively Christian.

[27]The "Son of God" is the *Logos*, who, as the expressed
intention of the Creator's Mind, informs and suffuses the
primal chaos of matter.  In the Hermetic cosmogony *Poimandres* (6) Mind is the seer's God (cf. Zosimos 12.9) and
the φωτεινὸς Λόγος born of Mind is Son of God.  Elsewhere
(*Corpus Hermeticum* 9.8 and 10.14) the world is called
"Son of God," but that cannot be the reference here.  The
role of the *Logos* in Hermetism derives from Stoicism, and
no influence from Christianity via the Prologue to the
Gospel of John, for example, nor from Judaism via a passage like Psalm 33:6 (32:6 LXX), need necessarily be as-
sumed, though it is not out of the question.

[28]Each individual soul.

[29]Or, perhaps, "to the Incorporeal One," i.e., God
who is Mind.

[30]The *Logos* is the messenger of Mind, sent to awaken
the kindred faculty in men.  Compare the role of Poimandres in *Poimandres* 2-3, and the result in 27f.  The idea
of the *Logos* as messenger of God in Hermetism may owe
something to the function of Hermes as ἄγγελος for the
court of Zeus in classical Greek mythology.

[31]That is, "man whose raw and inherent nature it is
to suffer, to be passive," a nature the very opposite of
that of Mind himself.  Compare *Corpus Hermeticum* 12.11
and note the parallel with Phil 2:6-7.  Although a com-
parable statement is made of Jesus Christ in 13:3-4 in
reference to the Passion (cf. 13.4), one need not follow
Reitzenstein in excluding this sentence as a Christian
interpolation.  There is nothing in it which is not in
accord with Hermetic theology.  Festugière, usually so
wisely cautious in this regard, marks καὶ πατρὶ ὑπακούει
in 7.17 out as Christian interpolation, thinking perhaps
of dependence upon Phil 2:8, but the conjunction of the
*Logos* with the Will of God in creating the world in *Poimandres* 8 argues against this supposition.

[32]Or, possibly, taking πάντα as subject and not ob-
ject of γίνεται, "all that he wills comes to pass."  So
Ruska; but this is less likely, as it does not accord with
the context here and conflicts with the similar statement
in 12.9-11.

[33]Each mind.

[34]Or, "where he (the Son of God) was before ever he
etc."

[35]Or translate passively "yearned after."

[36]Either the author of the Hermetic document Zosimos
is citing has already referred to a world of light (cf.
*Poimandres* 4-5, 6, 7) in an earlier section, or ἐκεῖνο may
simply look back to τὸν εὐδαίμονα χῶρον in 7.18-19. Compare
also 11.1 and 14.6-7.

[37]This Bitos is almost certainly identical with a
Bitus twice mentioned by the late Syrian Neo-Platonist
Iamblichus in his *On the Mysteries of the Egyptians*. In
the first passage (8.4) Iamblichus remarks about their
uranology that the Egyptians do not simply theorize, but
"recommend that you make the ascent, by means of the
sacred hieratic theurgy, through the higher and more universal regions that lie beyond Fate, to God the Creator,
and have neither anything to do with matter nor any concern for anything else, except only observance of the
appropriate moment" (cf. Zosimos 2.1 with note 9 and the
danger implied by 2.8-9). "Hermes," Iamblichus goes on
(8.5), "expounded this very way; the prophet Bitus found
it inscribed in hieroglyphic characters in shrines round
about Sais in Egypt, and translated it for king Ammon. He
handed on the name of God that extends through all the
world. And there exist," he adds, "many other (viz.,
other Hermetic) compositions on this subject." The Hermetic tracts from which Zosimos cites sections 4, 5, and
7.4ff., as well as *Corpus Hermeticum* 13, are examples of
just such other compositions. In the second passage, in
the context of discussion of the same "way"--liberation
from Fate by ascension and union with the supracosmic
God--Iamblichus adds (10.7): "What is Good-in-itself they
(the Egyptians) hold to be, on the divine level, the God
who exists beyond conception, and on the human level,
union with Him, which is what Bitus says, translating from
the Hermetic books." Bitos, or Bitus, if he existed,
would have been, like Manetho or Chairemon, a Hellenized
Egyptian priest and interpreter of native Egyptian traditions to the Greeks. The visionary author of the Hermetic
tractate from which Iamblichus derived his information
used Bitos' name as an authority to support the common
Hermetic claim of derivation from long-lost or previously
uninterpreted stelae of "Hermes," i.e., Thoth (the identification is implied already by Herodotus 2.138). Reitzenstein (*Poimandres*, pp. 106-8) entertains the possibility
that our Bitos may be identical with a certain Thessalian
king Pitus to whom three spells are ascribed by the great
Paris magical papyrus (*Papyri Graecae Magicae*, ed. Preisendanz, 4.1928, 2006, 2140) and with the Bithus of Dyr-
rhachium (in Illyria) mentioned by Pliny in his *Nat. Hist.*
28.23.82 for his recommendations on dealing with mirrors
tarnished by the look of a woman during menstruation.
Both identifications seem untenable to me. Reitzenstein
(*Poimandres*, p. 104, n. 1) also suggests comparison of the
tablet Zosimos claims Bitos to have written with the

"Ophite" diagram discussed by Origen in his *Contra Celsum* 6.30-32, and both Scott (*Hermetica* IV, p. 130) and Festugière (*La Révélation* I, p. 268, n. 1) assume that some sort of picture is meant here. But this need not be the case. Bitos "wrote" his tablet, and Iamblichus states that the prophet translated documents in hieroglyphic characters. Furthermore, Ruelle's emendation Κέβητος for the manuscripts' καὶ Βίτος, though unacceptable, is instructive for suggesting comparison with the *Tablet of Kebes*. That document poses as an interpretation of an ancient tablet with a strange scene depicted on it, but is not itself a picture. A πίναξ is, in any case, a writing board, and the word does not prejudge what is written upon it. It may be that "Tablet" was the title given the work written in Bitos' name--a title chosen to lend an air of hoary antiquity to the work.

[38] This passage is difficult because it is so elliptical. It seems to be saying: Look at the tablet that Bitos wrote, and (what) Plato and Hermes (wrote, and you will find) that etc. Festugière (*La Révélation* I, loc. cit.) takes up Reitzenstein's proposal (*Poimandres*, p. 104, n. 1) to assume that a chain of dependence is implied here, viz., Bitos upon Hermes, and Hermes upon Plato. The first link is certain, but the second is not. In Zosimos' mind the dependence was of Plato upon Hermes (equals Thoth), not the other way round; the epithets "thrice-great" and "infinitely-great" attached to Plato and Hermes, respectively, indicate the succession Zosimos presumes. The author of the Hermetic tract in Bitos' name may actually have depended upon Plato, at least partially and perhaps explicitly, for the statement in 8.3-5. In the *Phaedrus* 274Cf. Plato has Socrates report that he has heard that "one of the ancient gods," Theuth, discovered the use of numbers, geometry, astronomy, and most importantly, the alphabet; in the *Philebus* 18Bf. alphabetic innovations are credited to "a god or some god-like man" whom Egyptian tradition says was Theuth. "Bitos" may have derived his equation of Thoth with the first man from Plato's reference to a "god-like man," but his knowledge that such was his designation "in the original hieratic language" (8.3) came from "Hermes." See the preceding note. Alternately, the reference to Plato and Hermes may be an aside by Zosimos.

[39] The word "hieratic" (i.e., "sacred, priestly") refers, as the context here and the passage from Iamblichus in n. 37 show, to the most ancient form of the Egyptian language, the hieroglyphic pictographs, and not the script called hieratic derived from it. Egyptian is held to have been the first written language, because Thoth, as we saw, invented the alphabet.

[40] That is, the books containing this information, specifically, as what follows shows, the Septuagint, in particular of the book of Genesis, and allegorical exegesis

based upon it. See, for example, Josephus, *Jewish Antiq-uities* 1.34 and Philo, *Legum Allegoria* 1.90. That the first man was "the giver of names to all corporeal beings" is reminiscent of Gen 2:19. It is possible that the Hermetic Bitos-document was the source of all this section and perhaps the following one as well, and not Nikotheos or one of Zosimos' "Hebrew" sources; see notes 2 and 5 to the Introduction. The Hermetic cosmogony *Poimandres* (18), after all, cites Gen 1:28, and Jewish influence elsewhere in the Hermetica is common.

[41] Evidently Ptolemy II Philadelphus, by whom, accord-ing to legend (the *Letter of Aristeas* especially) the translation known as the Septuagint was commissioned.

[42] There is a play on words here impossible to pre-serve in English, namely, on the fact (noticed by Plato, for example: *Cratylus* 407E), that the name of the god Hermes is very similar to the verb ἑρμηνεύειν "to trans-late, to interpret." Hermes' Egyptian counterpart Thoth, a god of learning and skilled in speech, is described above as "the interpreter (ἑρμηνεύς) of all that exists" (8.4). See also 8.7 of Adam's name and 16.5 of Prometheus-Epimetheus. Festugière (*La Révélation* I, p. 268, n. 5) re-gards the manuscripts' Ἑρμῆν in 8.11 as impossible and adopts Scott's emendation ἑρμηνέα. Of course, in a purely Jewish context the reference to Hermes would be unthink-able, but if the source is Hermetic, the difficulty dis-appears. Hermes was the hermeneute *par excellence*. For an interesting parallel see Acts 14:11-12.

[43] In late Jewish circles Hebrew was, not unnaturally, held to be the language spoken by the angels; cf. 1 Cor 13:1 and 2 Cor 12:4.

[44] See note 5.

[45] Unfortunately for the scheme of initial letters in force here, the word "earth" in Greek does not begin with a delta; hence the jugglery.

[46] There must be a lacuna here, for four elements were mentioned in 9.4 and the explanation of the second A in the Greek form of Adam's name is missing. It must have involved correlation with the *arctic* north (ἄρκτος) and somehow with water, however an alpha might have been eked out for that element, since the Greek word for water (ὕδωρ) lacks the appropriate initial letter. Perhaps the scribe's eye jumped over the clause involving the second A, and it would thus only be coincidence that the Hebrew form of Adam's name lacks the very letter in question. The acros-tic puzzle worked out here is in any case possible only in Greek. See also *2 (Slavonic) Enoch* 30:8f.; Philo, *De Opi-ficio Mundi* 51; and the *Sibylline Oracles* 3.24-26.

[47]That is, the sun. The order of the planets
followed here is the moon, Mercury, Venus, the sun, Mars,
Jupiter, Saturn--the most common one.

[48]That is, the body. According to Gen 2:7 God
"moulded" (ἔπλασεν) Adam's body out of dust from the
earth, and Hephaistos "moulded" the body of Pandora "out
of earth" (Hesiod, *Works and Days* 70). Note also that
Plato, in his famous image of the soul as comprising a
man, a lion, and a many-headed beast (*Republic* IX, 588Bf.),
repeatedly uses this verb to describe the various stages
in the construction of his fantastic man, and then, at the
last stage: "Now mould around (περίπλασον) them (the lion
and the many-headed beast) on the outside the likeness of
one, that of the man."

[49]The inner Man, a purely spiritual being, naturally
has a "real, true" name as well as the "common" name that
tongues of flesh in our material world's language have
named him. Compare the distinction made, evidently by
Nikotheos, between an ἔνσωμος and an ἀσώματος φράσις in
section 1 with note 3 ad loc. Plato's distinction between
names embodied in sounds and αὐτὸ ἐκεῖνο ὃ ἔστιν ὄνομα the
"actual," "real," or "natural" name (*Cratylus* 389D-390A;
cf. also 390D-E) and between what men call things and what
gods call them (*Cratylus* 391D-E) may be the ultimate source
for the idea. The Naassenes, according to Hippolytus
(*Elenchos* 5.7.1f.), held that Kaulakau was Adamas, the
primal, heavenly Man, Saulasau the mortal one below, and
Zeesar the Jordan that flows upwards. The three names derive from Isaiah's mocking gibberish, Isa 28:10. "This,"
the Naassene account goes on, "is the hermaphrodite Man in
all man, whom the ignorant call Geryone the three-bodied"
(5.7.4). Compare also the Nag Hammadi *Gospel of Philip*
(codex II, 3: p. 56.3-4) and the obsession with knowing
the "authentic name" evident in the magical papyri.

[50]See 1.4 and note 4 on Nikotheos. Our Naassenes
(Hippolytus, *Elenchos* 5.7.2) spoke of "a kingless generation begotten above, where dwell Mariam the one who is
sought for (ἡ ζητουμένη) and Jethor the great sage and
Sepphora the one who sees and Moses, whose begetting was
not in Egypt." These legendary seers and prophetesses,
Moses and his family, are now, like Nikotheos, denizens of
heavenly realms as members of "the kingless generation."
Documents like the *Assumption of Moses* attest the early
development of this pattern. Jesus' "resurrection" and
continued inspiration of his followers by his spirit are
on the same prophetic trajectory; that the Son of Man,
like Enoch, or Moses, or Elijah, was "hidden"--in hiding--
and "could not be found" was a common motif in Jewish Messianic speculation. See also Gen 5:24 LXX of Enoch: "he
could not be found (οὐχ ηὑρίσκετο), because God translated
him." Deut 34:6 is the source of the legend in the case
of Moses: his grave could never be found.

[51]There is once again a play on words here impossible to preserve in English. A common Homeric term for man, φώς, is held to derive from the word φῶς, contracted from φάος, "light." The derivation is significant for Gnosticism since the primal Man is commonly considered to be light as originating from the world of light, but etymologically the two words are unrelated. See also 11.1; 13.2, 5, 12; and 14.6.

[52]Zosimos, or his Nikotheos-source, is playing upon the well-known triple significance of πνεῦμα, "wind," "breath," and "spirit." In the Garden of Eden Phos, a πνευματικὸς ἄνθρωπος (10.2-3), rides the wind whose nature is akin to his own. Spirits and unembodied souls dwelt in the air. Compare, for example, the complaint of souls about to be embodied in the 23rd Hermetic extract from Stobaeus (the *Kore Kosmou* 36): "What a wretched thing it will be for us to have to bear hearing our kindred spirits blowing (or, breathing: φυσώντων) in the air!" The idea stems from Orphic conceptions of the soul; Aristotle (*De Anima* 410[b]19) reports that in Orphic poems Orpheus is represented as teaching that "the soul of beings that breathe enters them from the universe borne along on the winds." The fact that in Gen 2:7 LXX God is said to have "breathed" or "blown into" (ἐνεφύσησεν) Adam's face "a breath of life" would, to one familiar with Hellenistic philosophies, stimulate such speculation, especially since the Gnostics interpreted Gen 3:21 of the embodiment. Compare also John 3:8 τὸ πνεῦμα ὅπου θέλει πνεῖ in its context, and Jesus' method of conferring the spirit upon his disciples in 20:22.

[53]The only conceivable antecedent is τὰ στοιχεῖα, but here the "elements" are the four physical elements out of which Phos' body is composed (11.4). They must, as Reitzenstein saw, be the archons, the planetary rulers who are the agents of Fate. Perhaps Zosimos' Nikotheos-source mentioned them explicitly in a passage preceding that which Zosimos is citing.

[54]The term ἀνενέργητος means "without ἐνέργεια" in the Aristotelian sense of that word, that is, "without expression in act," "unactivated," "unactualized." Here the meaning would be that Phos, before his embodiment, existed in a purely potential state and lacked the modes of self-expression proper to the planetary rulers and their body derived from the elements. In the *Poimandres* 24-25 the soul, in its reascent through the spheres at death, leaves behind it, "inactive" (ἀνενέργητον) now, every gross and vicious state of being with which the planets had clothed it in its descent through them into birth.

[55]Hesiod (*Theogony* 521-522) says that Zeus "bound (δῆσε) wily-minded Prometheus with galling (or, possibly,

inextricable) bonds, grievous fetters (δεσμοῖς)" in reference to the Titan's crucifixion upon Caucasus. Pandora
is never called a "fetter." After the account of her
creation in 570ff., however, there is another reference
to Prometheus' bondage. Zeus "confined him under con-
straint (ὑπ' ἀνάγκης: cf. Zosimos 11.1-2) with a great
fetter" (615-616). This could be understood to refer to
Pandora.

[56]Prometheus' name means "Forethinker"; Aeschylus
(cf., for example, *Prometheus Bound* 85-87) already plays
upon its connection with προμήθεια "forethought," foresight." Because of his brother Epimetheus' ("After-
thinker's") disobedience of his advice in the matter of
Pandora Prometheus falls into bondage to fleshly existence.
But his fate serves as an exemplar to those that possess
mind; see 16.3-5. The allegorization of the two brothers
as two facets of the single man, hindsighted flesh and
foresighted spirit, follows naturally from the significance of their names.

[57]See 7.12f. and notes 27 and 30.

[58]Or "reveals himself to each (viz., individual
mind)." For another example of the use of an active form
of φαίνω with middle meaning, see *Papyri Graecae Magicae*
(ed. Preisendanz, 1.90), φαῖνέ μοι, θεέ. The more normal
middle form occurs in 13.2 from another source.

[59]Or even "joined himself."

[60]That is, back to Paradise. The first sentence of
section 13 represents a form of the legend, best known
from the version embedded in the *Acts of Pilate* (see
Hennecke-Schneemelcher-Wilson, *New Testament Apocrypha* I,
pp. 470ff.), of the rescue of Adam by the triumphant
Christ in his *descensus ad inferos*. The original context
for this late and watered-down orthodox version is the
late Jewish soteriological Adam-speculation and ascen-
sional mysticism that developed during the course of the
Hellenistic period (compare the role of Sophia as libera-
tor of Adam in Wis 10:1-2) and deeply affected incipient
Gnosticism, as well as Paul (see Rom 5:12-21, 1 Cor-
inthians 15, especially 20-28 and 44-57, and 2 Cor
12:1-6--his opponents boasted of the same experiences).
In our Nikotheos document the Hades is our material world
tyrannized by Fate and Adam is Phos, the inner Adam
trapped in the body (note Paul's reluctance to specify
whether his vision took place in or out of the body),
not the corporeal Adam. The contrast of Adam to "very
powerless men" in 13.3 makes this a certainty. Eric
Peterson in a very interesting article ("La Liberation
d'Adam de l''Ανάγκη" in *Revue Biblique* 55 [1948] pp.
199f.), shows that the Jewish prayer preserved in two
recensions in Preisendanz' collection of magical papyri

(1.195f. and 4.1167f.) and the so-called "Mithras-Liturgy" (4.475f.: see now the edition by Marvin Meyer in this same Texts and Translations series) as well, stem from the same background. In the Mimaut papyrus (*Papyri Graecae Magicae* 3.144) the reference to Adam is explicit. Reitzenstein, Festugière, and Schenke all mark out much of this section as Christian interpolation, but, if there is any at all, it is better accounted for as a secondary christianiza- tion of the Nikotheos document itself by Christian Gnostics--a process to which the Nag Hammadi tractates (*Eugnostos the Blessed* and the *Sophia of Jesus Christ*, for example) have made us more sensitive--than as a later Byzantine Christian interpretation of Zosimos' work, as Reitzenstein (*Poimandres*, p. 105, n. 4) holds. The doce- tic character of the Christology of this section and the integrity of the following one both tell against such a supposition.

[61] The inner light-men, particles of Phos. See n. 51.

[62] Before their embodiment. In the Hermetic version in 7.18f. Nous plays a personal (cf. "our Mind" in 12.9), salvific role analogous to that of Jesus Christ here. See also 14.6-8.

[63] See 7.16 and n. 31. Here, however, παθητός is not a general description, but refers specifically to the Passion, as ῥαπιζόμενος shows: compare Mark 14:65 "the servants struck him blows" and 15:19 and parallels. In the *Acts of John* 96 Jesus tells his disciples: "Yours is this passion of the Man which I am about to suffer." The Passion was real, but only for the corporeal Jesus; the immaterial Jesus, the spiritual Man, suffered nothing (Zosimos 13.5). In the *Acts of John* John witnesses the crucifixion and runs away, but in a cave on the Mount of Olives hears the truth from his Lord's disembodied voice (97). Compare also the grossly docetic Christology of- fered in 93.

[64] I have reluctantly accepted Reitzenstein's emenda- tion ἔπεισι for the manuscripts' τόποισι, reluctantly be- cause τόπος is a technical term in Gnosticism for the "re- gions" of the Demiurge's cosmos, or for the Demiurge him- self: see, for example, Clement of Alexandria's *Extracts from Theodotus* 34.1-2; also 37.1 and 4.2. Such a refer- ence would be quite apposite here, but I cannot work it in; the context seems to demand a verb.

[65] Literally "to make exchange for." καταλλαγή cannot mean "reconciliation" here as it does in the New Testament.

[66] The text has simply an indefinite παρ' αὐτῶν in both cases. The reference must be back to 11.3 where the same prepositional phrase occurs, that is, to the planetary archons. It is possible, but less likely, that παρ' αὐτῶν is equivalent to παρ' αὐτοῖς and means "that they (the "photes," the inner men) bear with them."

[67]In Gnostic systems the evil Demiurge consistently imitates the patterns of the world beyond him. The root idea of imitation is Platonic, but for the Gnostics the Demiurge is not, as for Plato, good, but bent upon deception and enslavement. In the *Apocryphon of John* (I cite the long recension of Nag Hammadi codex II, 1 in Søren Giversen's edition *Apocryphon Johannis* [Copenhagen: Munksgaard, 1963]), the "imitated spirit" stands in opposition to the "spirit of life" and leads the souls of men astray (74.7-75.11). Ialdabaoth creates Fate in jealousy of the world of light that surpasses him in knowledge and glory to deceive and enslave the cosmos to ignorance and oblivion. He seeks to destroy Noah and the whole generation "that does not waver" with a flood, but fails, and so sends his angels to the daughters of men to sow their evil seed in them. Failing again in their attempt, they create "a spirit which imitates the image of the spirit (of life) which came down (from the world above) in order that the souls through this might be defiled, and the angels changed themselves in their likeness according to the image of their (the women's) husbands, and filled them with the spirit of darkness and wickedness which they mingled in them....They took women and they engendered from the darkness sons after the image of their spirit, and they closed their hearts and they were hardened through the hardness from the imitated spirit until now" (75.31-78.11). From this passage it is obvious that this speculation about the evil, imitating spirit is a development of the post-Persian period Jewish concept of an "evil inclination" (יֵצֶר רָע) and a "good inclination" (יֵצֶר טוֹב) struggling for domination in the hearts of men. See, for example, Sir 33:13-14; *Testament of Gad* 5:3, *Asher* 1:3, and *Benjamin* 6:1f.; *Damascus Document* 2.14f.; *Bereshith Rabba* 27. In Judaism, however, *both* spirits are created by God--a problematic view, to say the least. The *Apocryphon of John*'s connection of the creation of the counterfeit spirit with Noah's generation is not fortuitous, but points up the original Jewish background of the idea, statements in Gen 6:5 and 8:21 that the "inclinations" or "intentions" in men's hearts were always evil. The *Pistis Sophia* preserves the Greek form ἀντίμιμον πνεῦμα. There, as in the *Apocryphon of John* upon which it is probably dependent, the "mimicking spirit" is an instrument of Fate in man (cf. Zosimos 14.10-11); its purpose is to awaken sinful desires (chaps. 111-112). It accrues to the soul by making itself a perfect resemblance (131).

[68]In the Garden as the Serpent. The Serpent mimicked God (Gen 3:1 as against 2:16-17) to mislead Eve and so brought about Adam's fall. For the Antichrist as Deceiver see especially 2 John 7.

[69]See note 51.

[70]Either "before (the foundation of) the world" or "before (their coming into) the world," which amounts to the same. See 11.1 and 7.19-20.

[71]That is, before he comes claiming to be Son of God.

[72]Reitzenstein and Festugière plausibly suggest Μανιχαῖος as the solution to the riddle. But why this caustic polemic against Mani? The explanation may be, in addition to the reason suggested in the Introduction (p. 5), that the community in which the Nikotheos document originated was connected somehow, if not identical, with the Babylonian Jewish-Christian baptist sect in which Mani had been raised and with whose members he suffered a violent breach over the revelations that formed Mani's independent calling. The Cologne Mani-Codex has revealed that this sect was Elchasaite (see Albert Henrichs' article "Mani and the Babylonian Baptists: A Historical Confrontation" in the *Harvard Studies in Classical Philology* 77 [1973] pp. 23ff.), and Elchasai has much in common with Nikotheos. Broadly speaking, both were visionaries and apocalyptists (for Elchasi see fragment 7 from Hippolytus, *Elenchos* 9.16.2-4 in Hennecke-Schneemelcher-Wilson, *New Testament Apocrypha* II, pp. 745f.); more specifically, both concerned themselves with escape from the evil influence of stars and Fate and communicated mysteries that could not be "interpreted" (Elchasai fragment 9 from Epiphanius' *Panarion* 19.4.3). Elchasai's teachings were carried to Rome by a certain Alcibiades, a native of Apamea in Syria; they had, he claimed, been delivered to Elchasai by a pair of gigantic angels, the Son of God and the Holy Spirit (fragment 1 from Hippolytus 9.13.1-3). The Nikotheos document might, then, have been a product of this evangelization of the west. The author of the Johannine Apocalypse was roughly Elchasai's contemporary--Elchasai began to preach in the third year of Trajan's reign (A.D. 101) according to fragment 2 (Hippolytus 9.13.3)--and perhaps Nikotheos' also. He, like them, was a seer and apocalyptist in the Jewish-Christian tradition. Compare the riddle here with that on the name of the Beast in the Apocalypse 13:18. Mani did indeed make great claims for himself, to be the Παράκλητος promised by Jesus in John 14:16-17 and 26, for instance (see Hegemonius, *Acta Archelai* 34), and the fantastic Manichaean cosmological system could easily be characterized as μυθόπλανος by an opponent.

[73]The diphthong could be counted as one letter instead of two, especially since diphthongs were often written that way (-ε for ει, to choose but one example).

[74]That is, "matching the pattern of the word 'fate'" which has nine letters in Greek (εἱμαρμένη), including a diphthong. The author may, however, also be continuing his effort to connect the deceptive teachings of his opponent Mani with the operation of Fate (14.10-11).

[75]This must be the Mimic *Daimon*, not the Son of God; see the Introduction, p. 5 and note 21 there. The Antichrist's forerunner comes first and initiates the period of "Messianic woes" (compare Mark 13:5-6, 14, 21-22 and parallels, and the Johannine Apocalypse 13), after which the Evil One comes himself, openly, with the same monstrous claim and purpose (see 2 Thess 2:3-12, especially 7-8, and the Apocalypse 20:7-8). The seven periods probably refer to the Iranian conception of the progress of creation in millenia of struggle between Good and Evil toward the final triumph of the former; see Plutarch, *De Iside et Osiride* 47 (370B-C) and the Pahlavi *Bundahish* 1.20. The ugliness of the Evil One in body as well as soul (14.3-4) is similarly Iranian (*Bundahish* 1.11 and 3.9 where his body is a lizard's).

[76]See note 2 to the Introduction.

[77]Prometheus, not Epimetheus. It is Prometheus, the spiritual man, who suffers bondage in the flesh for the stupidity of his mortal brother. See 12.1-9. No lacuna is necessary in 16.5.

[78]Or, "repented," but μετανοήσας plays upon the fact that Prometheus is νοῦς and signifies not so much repentance here as a new orientation of mind, a concentration upon its original state that involves it in disenfranchisement from matter and a return to the "realm of bliss" beyond Fate.

SELECT BIBLIOGRAPHY


Berthelot, Marcellin. *Histoire des Sciences: la Chimie au Moyen Âge*. 3 vols. Paris: Imprimerie nationale, 1893.

_____, and Ruelle, Charles-Émile. *Collection des Anciens Alchimistes Grecs*. 3 vols. Paris: G. Steinheil, 1887-1888. Reprinted in one volume; London: Holland Press, 1963.

Bidez, Joseph, and Cumont, Franz. *Les Mages Hellénisés*. 2 vols. Paris: Société d'éditions "Les Belles Lettres," 1938. Reprinted in one volume; New York: Arno Press, 1975.

Bousset, Wilhelm. *Hauptprobleme der Gnosis*. Göttingen: Vandenhoeck und Ruprecht, 1907.

Ferguson, Alexander, and Scott, Walter. *Hermetica*. 4 vols. Oxford: the Clarendon Press, 1924-1936.

Festugière, André-Jean. "Alchymica." *L'Antiquité Classique* 8 (1939) 71-95.

_____. "La Doctrine des 'Viri Novi' sur l'Origine et le Sort des Ames." Pp. 97-132 in *Mémorial Lagrange* (Cinquantenaire de l'École biblique et archéologique française de Jérusalem). Paris: J. Gabalda, 1940.

_____. *La Révélation d'Hermès Trismégiste*. 4 vols. Paris: J. Gabalda, 1950-1954.

Forbes, Robert. "Chemie." Cols. 1061-1073 in *Reallexikon für Antike und Christentum* II. Stuttgart: Hiersemann Verlag, 1954.

_____. *Studies in Ancient Technology* I. Leiden: E. J. Brill, [2]1964, pp. 125-148.

Jung, Carl. *Psychologie und Alchemie*. Zürich: Rascher Verlag, [2]1952. Eng. trans. by R. Hull. *Psychology and Alchemy* (Bollingen Series 20). Princeton, NJ: Princeton University Press, [2]1968.

Nock, Arthur Darby, and Festugière, André-Jean. *Corpus Hermeticum*. 4 vols. Paris: Société d'éditions "Les Belles Lettres," 1946, 1954.

Quispel, Gilles. "Der gnostische Anthropos und die jüdische Tradition." *Eranos-Jahrbuch* 12 (1954) 195-234. Reprinted in *Gnostic Studies* I, pp. 173-195. Istanbul: Nederlands Historisch-Archaeologisch Institut in het Nabije Oosten, 1974.

Reitzenstein, Richard. *Poimandres*. Leipzig: B. G. Teub-
    ner, 1904.

Riess, Ernst. "Alchemie." Cols. 1338-55 in Pauly-
    Wissowa, *Realencyclopädie der classischen Altertums-
    wissenschaft* I. Stuttgart: J. B. Metzlerscher Verlag,
    1893.

Ruska, Julius. *Tabula Smaragdina*. Heidelberg: C. Winter,
    1926.

Schenke, Hans-Martin. *Der Gott "Mensch" in der Gnosis*.
    Göttingen: Vandenhoeck und Ruprecht, 1962.

# CODICES

M     Venetus Marcianus 299  saec. X-XI (Berthelot)
vel XI (Zanetti) vel XII (Morelli): cf. J.
Bidez, F. Cumont, A. Delatte, J. Heiberg,
et O. Lagercrantz, *Catalogue des Manuscrits
Alchimiques Grecs* II: Les Manuscrits Italiens
(Bruxelles: M. Lamertin, 1927) pp. 1ff.
f. 189$^r$ - f. 193$^r$

K     Parisinus graecus 2249  saec. XVI
f. 97$^r$ - f. 100$^v$

Νικόθεος 1.4; 10.5
νοερός 16.6
νοέω 7.12
νομίζω 7.7
νοῦς, Νοῦς 7.18; 12.7, 9;
   13.9; 16.2

ὁδηγέω 7.21
ὁδηγός 15.3, 4
'Ολύμπιος 6.5
ὁμολογέω 2.13; 16.8
ὁμολογία 3.5
ὄνομα 10.3, 6; 14.12
ὀνοματοποιός 8.4
ὄντα, τὰ 8.4
ὄργανον 1.8; 2.5
ὀρέγω 7.20
ὅρος 14.13
ὅσιος 7.13

παθητός 7.16; 13.4
παιδευτήριον 4.5-6
Πανδώρα 12.3
παραγγέλλω 6.2
παράδεισος 11.1
παρακαλέω 8.10
παρακοή 12.8
παρακολουθέω 10.7
παρακούω 12.8
παραλαμβάνω 2.9
παρθένος 8.7
Πάρθος 8.5
πάσχω 13.5
πατήρ 7.17
πείθω 2.7; 11.2
πέμπω 8.11
πεπαντικός 9.9
πέρας 5.6
περιέχω 1.8
περίοδος 14.14
περίπλασις 10.2
Περσίς 14.10
πηλός 7.11
πίναξ 8.1
πλανάω 14.3
πλάνη 15.5
Πλάτων 8.2
πλοῦτος 6.4
πνεῦμα 14.6
πνευματικός 7.5; 10.3; 13.11
πολιτεύω 7.12
πομπή 4.3
πορεύω 7.8
πρᾶγμα 3.6

πρόδρομος 14.9
Προμηθεύς 6.1; 12.2, 4, 7,
   8-9
πρόσειμι 13.1
προσηγορικός 10.4, 6
Πτολεμαῖος 8.9
πῦρ 9.9
πυρρός 8.7

ῥαπίζω 13.4
ῥῆμα 2.12

Σαραπεῖον 8.10
σάρκινος 8.8; 10.1
σάρξ 12.7
σπορά 1.6
στοιχεῖον 1.1, 8; 9.4, 5,
   6, 8; 11.4; 14.11-12
στρογγύλος 1.1
συλάω 13.5
συμβουλεύω 13.8; 16.2, 5
συμβολικός 9.3
σύνειμι 13.8
σφαῖρα 9.4
σφάλλω 16.4
σχηματίζω 3.3
σῴζω 14.7, 12-13
σῶμα 7.11, 17; 9.5, 9;
   12.6; 14.4
σωματικός 3.4, 8; 4.5; 7.19;
   8.5; 16.6

τέλος 13.7
τέχνη 2.10
τολμάω 14.8
τρέπω 2.11
τρίας 7.10
τρίσμεγας 8.2
τυφληγορέω 13.11
τύχη 2.10

ὕδωρ 1.8
υἱός 7.13; 12.9; 14.3, 5;
   15.3, 5
ὑπακούω 7.17

φαίνω 10.1-2; 12.11; 13.2
φανερός 13.8
φαντάζω 3.8-9; 4.3, 7
φάσκω 7.2
φιλοσοφία 6.7

www.ingramcontent.com/pod-product-compliance
Lightning Source LLC
Chambersburg PA
CBHW031933080426
42734CB00007B/671